GEOMETRY

Lessons for Self-Study with Test Preparation

Build Your Self-Confidence and Enjoyment of Math!

Thoroughly explained concepts and detailed proofs

with a Comprehensive Solutions Manual

Aejeong Kang

MathRadar

Send all inquiries to:

MathRadar, LLC
5705 Spring Hill Dr.
Mckinney, Texas 75072

Visit www.mathradar.com for more information and a sneak preview of the MathRadar series of math books.

Send inquires via email at info@mathradar.com

Geometry

ISBN-13: 978-0-9893689-3-3

ISBN-10: 0989368939

Printed in the United States of America.

Preface

I wrote these books because I am a mother and I have a strong academic background in mathematics. I have a BS degree in Mathematics and Master's degree in Mathematics as well. I have completed Ph.D. program in Biostatistics.

After receiving the big blessing of our first child, a daughter, I decided to forgo my personal career goals to become a full-time mother. When our daughter entered 7th grade, that meant lots of help with her study of math-my passion. However, I struggled to find good math books that would help her understand difficult concepts both clearly and quickly. After the conversation with my husband and (now two) children, I decided that the best way to help my children was by writing math books for them myself. They wholeheartedly agreed.

That's why I've been able to pour all my knowledge, energy, and soul into these books. Because I'm a mom, I would do anything for my children. Thanks to my family's endless support, I wrote them four books, designed for use in junior high and high-school (partially) mathematics.

And that would have been the end of my journey, but my husband and children insisted that I share my work outside of our family. They encouraged me to make my work available to other parents looking, as I was, for well-written, great mathematics books for their children.

So I finally decided to publish these books. I do so with the hope that they will help your children find success and confidence in learning and studying mathematics.

But I would never have begun or finished this project without the support of my family. Kyungwan, Nichole, and Richard, you are my world. Thank you.

Introduction

✔ *After reading several pages of explanation/description about a certain mathematical concept, you still don't get it.*

✔ *You have worked on many related problems to understand mathematical concepts, but you still feel completely lost in the mathematical jungle.*

✔ *You bought a math book with good reviews, but it only offers short answers without detailed solutions. You feel confused and frustrated.*

✔ *You've tried multiple learning math books, but you've still not getting good grades in math. It seems like math is just not for you.*

If any one of these situation sound familiar, the MathRadar series will help you escape!

Everyone has different learning abilities and academic skill. The MathRadar series is written and organized with emphasis on helping each individual study mathematics at his/her own pace.

In the case of Algebra, each book covers all the topics required in each field of Algebra. These fields were systematically subdivided into Part I, Part II, and Part III.

Algebra, Part I covers information about **Number Systems** from natural numbers to real numbers.

Level I	for **grades 6~8**	: Chapter 1 and Chapter 2
Level II	for **grades 7~9**	: Chapter 3
Level III	for **grades 8~10**	: Chapter 4

Algebra, Part II covers information about **Expressions** dealing with equations and inequalities.

Level I	for **grades 6~8**	: Chapter 1 and Chapter 2
Level II	for **grades 7~9**	: Chapter 3, Chapter 4, and Chapter 5
Level III	for **grades 8~10**	: Chapter 6, Chapter 7, and Chapter 8

Algebra, Part III covers information about <u>**Functions**</u>, while also including <u>**Statistics and Probability**</u>.

Level I	for **grades 6~8**	: Chapter 1	**Statistics** : Chapter 1, Chapter 2, and Chapter 3
Level II	for **grades 7~9**	: Chapter 2	**Probability** : Chapter 4
Level III	for **grades 8~10**	: Chapter 3	

Each book consists of clean and concise summaries, callouts, additional supporting explanations, quick reminders and/or shortcuts to facilitate better understanding.

With the numerous examples and exercises, students can check their comprehension levels with both basic and more advanced problems.

Each book includes <u>**Solutions Manual.**</u> The solutions manual makes it possible for students to study difficult concepts on their own. With the solutions manual, students will be able to better understand how to solve problems through step-by-step for each problem.

Geometry has also been systematically subdivided so that students can easily grasp geometry concepts. Each concept is thoroughly explained with step-by-step instruction and detailed proofs.

Carry the MathRadar series with you!

Work on them anytime and anywhere!

Finally, you can start to enjoy mathematics!

Whether you are struggling or advanced in your math skills, the MathRadar series books will build your self-confidence and enjoyment of math.

I hope Math Radar is what you need and will be a great tool for your hard work.

Your comments or suggestions are greatly appreciated.

Please visit my website at www. mathradar.com or email me at aejeong@mathradar.com

Thank you very much. And remember, math can be fun!

Aejeong Kang

TABLE OF CONTENTS

Chapter 2   Triangles and Polygons

Chapter 3 Quadrilaterals

Chapter 4   Circles

Chapter 5　Geometric Constructions

Chapter 6   Polygonal Regions and their Areas

Chapter 7   Solids and their Volumes; Surface Areas

Chapter 8 Similarity

Chapter 9   The Pythagorean Theorem

Chapter 10   Trigonometric Ratios

Solutions Manual

Index

Basic Geometry

Chapter 1 Basic Geometry

1-1 Number Lines and Distance

1. Number Lines

2. Distance

    (1) Absolute Value

    (2) Distance

    (3) Coordinates

1-2 Lines and Midpoints

1. Lines, Segments, and Rays

    (1) Lines

    (2) Segments

    (3) Rays

2. Midpoints

1-3 Planes

1. Defining a Plane

2. Relationships

    (1) Points, Lines, and a Plane

        1) Two Points and a Line

        2) Points and a plane

        3) A Point and a Line

    (2) Lines and Planes

        1) Two Lines

        2) Two Lines and a Plane

        3) A Line and a Plane

        4) Two Planes

3. Perpendicular Lines and Planes

    (1) The Perpendicular Bisector

    (2) The Perpendicular Segment

1-4 Angular Measures

1. Angles

    (1) Angle, Side, and Vertex

    (2) Classifying Angles

        1) Acute Angles

        2) Right Angles

        3) Obtuse Angles

        4) Straight Angles

    (3) Supplementary and Complementary

    (4) Congruence

    (5) Vertical Angles

2. Angles formed by a Transversal

    (1) Transversals

    (2) Corresponding Angles and
Alternate Interior Angles

    (3) Angles and Parallel Lines

CHAPTER
1

Chapter 1. Basic Geometry

1-1　Number Lines and Distance

1.　Number Lines

In a line, the counting numbers are arranged from left to right. We call them *positive integers*.

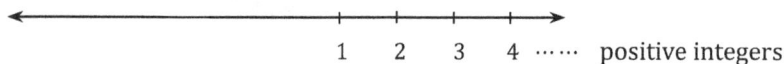

1　2　3　4　······　positive integers

Arranging the number 0 and negative integers (positive integers with negative signs in front) on the line from right to left, we have a *number line*.

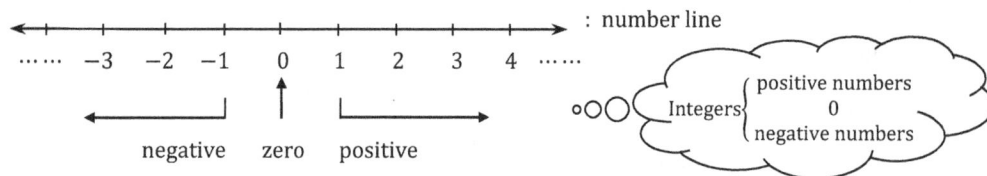

: number line

······　−3　−2　−1　0　1　2　3　4　······

negative　zero　positive

Integers { positive numbers
0
negative numbers }

To fill a number line, we also include *rational numbers* that can be written as integers divided by other non-zero integers.

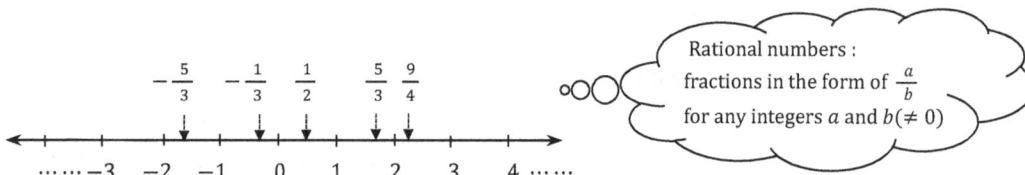

$-\dfrac{5}{3}$　$-\dfrac{1}{3}$　$\dfrac{1}{2}$　$\dfrac{5}{3}$　$\dfrac{9}{4}$

······−3　−2　−1　0　1　2　3　4　······

Rational numbers :
fractions in the form of $\dfrac{a}{b}$
for any integers a and $b(\neq 0)$

There are also many other numbers which cannot be expressed as fractions. We call them *irrational numbers*.

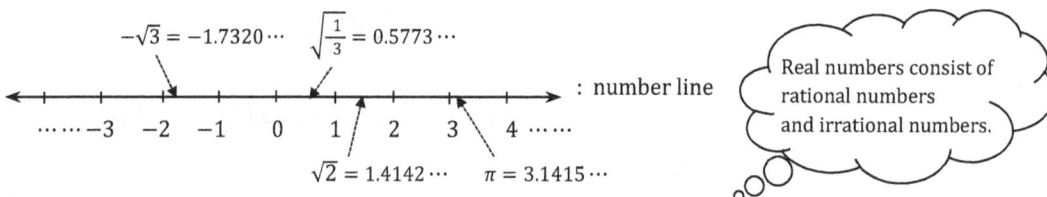

$-\sqrt{3} = -1.7320\cdots$　$\sqrt{\dfrac{1}{3}} = 0.5773\cdots$

······−3　−2　−1　0　1　2　3　4　······　: number line

$\sqrt{2} = 1.4142\cdots$　$\pi = 3.1415\cdots$

Real numbers consist of
rational numbers
and irrational numbers.

Every number on a number line has a point assigned to it.

All numbers that can be attached to points on a number line are called *real numbers*.

2. Distance

(1) Absolute Value

For any number a, the *absolute value* of a is denoted by

$$|a| = \begin{cases} a, & a \geq 0 \\ -a, & a < 0 \end{cases}$$

For example, $|0| = 0$, $|2| = 2$, $|-2| = -(-2) = 2$

Absolute value is always greater than or equal to 0 ($|a| \geq 0$).

(2) Distance

For any two points A and B on a number line, the *distance* between A and B is always a positive number. Therefore, we use absolute value to express the length.

The distance between the end points A and B is defined by a real number :

$$AB = |A - B| = |B - A|$$

If $A = B$ (A and B are the same point), then $AB = 0$.

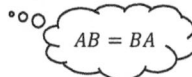

$AB = BA$

(3) Coordinates

The number associated with a point is called the *coordinate* of the point.

For example, in the number line below,

the coordinate of A is -2, the coordinate of B is 0 and the coordinate of C is 1.

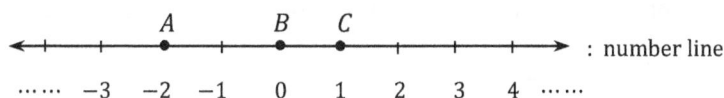

The distance AB is $|-2 - 0| = |0 - (-2)| = 2$ and the distance AC is $|-2 - 1| = |1 - (-2)| = 3$.

1-2　Lines and Midpoints

1. Lines, Segments, and Rays

(1) Lines

For any two different points A and B, the *line AB* contains A and B and is denoted by $\overleftrightarrow{AB}$.

While there are an infinite number of lines passing through a single point, there exists exactly one line passing through two different points.

point :　　　　　　　　　　line AB $\left(\overleftrightarrow{AB}\right)$:

(2) Segments

For any two different points A and B, the *segment* $\overline{AB}$ is a set with the end points A and B (in other words, the part of the line AB from point A to point B).

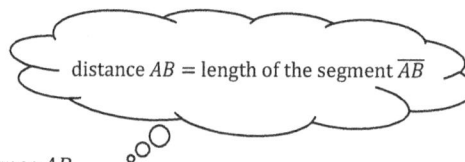

segment $\overline{AB}$:

distance AB = length of the segment $\overline{AB}$

Note :　The segment $\overline{AB}$ is different from the distance AB.

∵ Since the distance AB is $|A - B|$ which is a number measuring the length between the end points,

the distance AB is the length of the segment $\overline{AB}$.

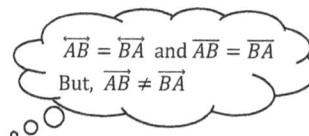

$\overleftrightarrow{AB} = \overleftrightarrow{BA}$ and $\overline{AB} = \overline{BA}$
But, $\overrightarrow{AB} \neq \overrightarrow{BA}$

(3) Rays

Starting from a point A, the *ray* $\overrightarrow{AB}$ proceeds through a point B in a straight line and in the same direction.

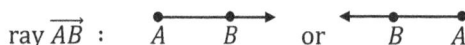

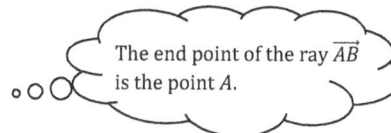

The end point of the ray $\overrightarrow{AB}$ is the point A.

ray $\overrightarrow{AB}$:　A　B　or　B　A

Note : If there is a point A between the points B and C, ray $\overrightarrow{AB}$ and ray $\overrightarrow{AC}$ are opposite.

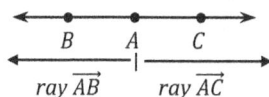

B　A　C

ray $\overrightarrow{AB}$　　*ray $\overrightarrow{AC}$*

2. Midpoints

The distance AB is the same as the length of a segment $\overline{AB}$.

The *midpoint* of a segment $\overline{AB}$ is the point which cuts the distance AB in half.

If M is the midpoint of a segment $\overline{AB}$, $AM = MB$.

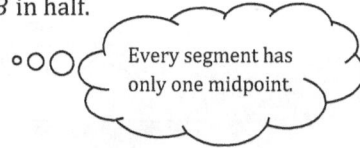

Every segment has only one midpoint.

Example

If the coordinate of point A is -6 and the coordinate of point B is 2 on a number line,

then the coordinate of the midpoint M is -2.

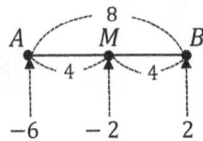

($\because$ Since $AB = |A - B| = |B - A| = 8$ and $AM = MB = \dfrac{8}{2} = 4$,

the point M is placed 4 units to the right of Point A along the line.

So, the coordinate of midpoint M is $-6 + 4 = -2$.)

1-3   Planes

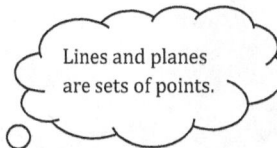

Lines and planes are sets of points.

1. Defining a Plane

A set of points lie on the same straight line is called a *collinear*.

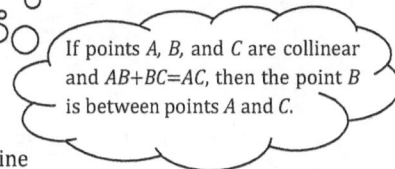

If points A, B, and C are collinear and $AB+BC=AC$, then the point B is between points A and C.

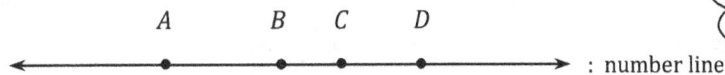

$\Rightarrow$ Points $A, B, C,$ and D form a collinear set.

To create a plane, we need at least three non-collinear points.

Case 1. A plane exists when there are at least three distinct points which do not pass through a line.

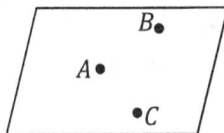

Case2. A plane exists when there are two distinct points on a line and a third point which does not belong to the line.

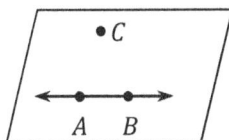

Case3. A plane exists when there are two lines which intersect at one point.

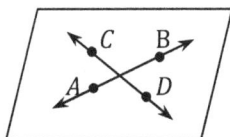

Case4. A plane exists when there are two parallel lines (lines with no intersection).

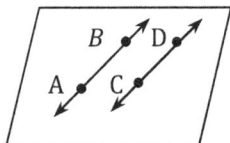

Note : If two distinct planes intersect, then their intersection is a line.

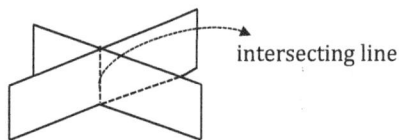

intersecting line

In order to fine an intersecting line between two distinct planes, the planes should not be parallel.

2. Relationships

(1) Points, Lines, and a Plane

1) Two Points and a Line

For any two distinct points, there is only one line which contains both points.

$A \qquad B$

2) Points and a plane

Every plane contains at least three non-collinear points.

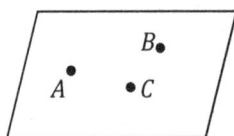

3) A Point and a Line

For a point P and a line $\overleftrightarrow{AB}$, there are two possible cases :

① the point is on the line, or

② the point is not on the line.

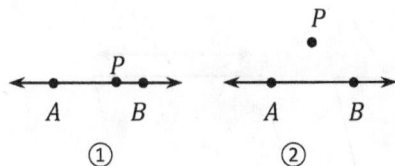

(2) Lines and Planes

1) Two Lines

For two lines, there are three possible cases :

① the lines intersect at one point,

② the lines do not have any intersection, or

③ the lines are congruent (have infinite intersections).

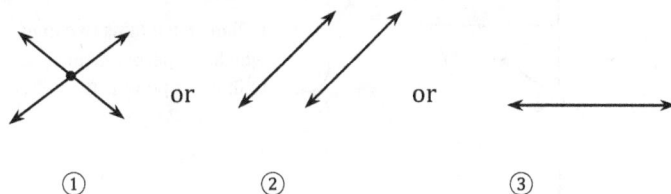

2) Two Lines and a Plane

For two lines and a plane, there are two possible cases :

① the plane contains both lines, or

② the plane does not contain both lines (one line intersects the plane rather than being a part of it). In this case, there is only one intersection point between the line and the plane.

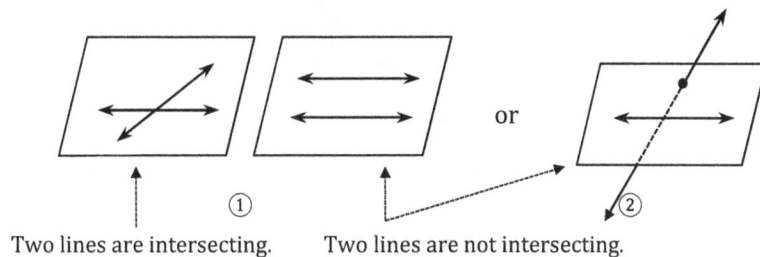

Two lines are intersecting.   Two lines are not intersecting.

3) A Line and a Plane

For a line and a plane, there are three possible cases :

① the plane contains the line,

② the plane and the line intersect at one point, or

③ the plane and the line are parallel.

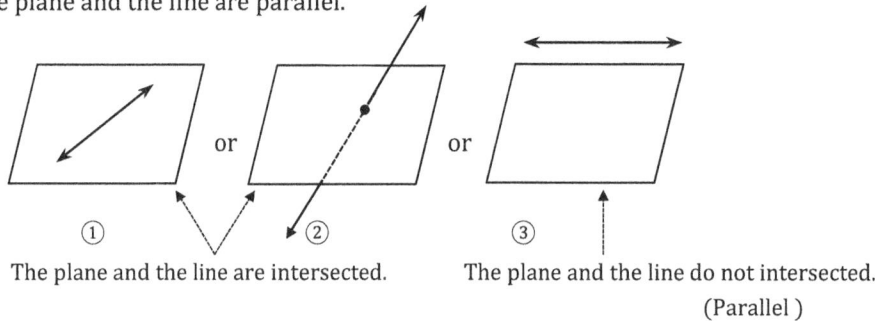

The plane and the line are intersected.   The plane and the line do not intersected.

(Parallel)

Note : Line l and plane E are perpendicular (l ⊥ E)

if   ① they intersect at one point P and

② all lines lying in the plane and passing through the intersection point are perpendicular to the

given line l.

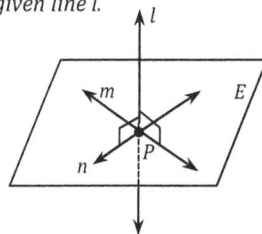

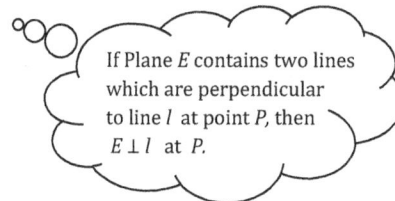

If Plane *E* contains two lines which are perpendicular to line *l* at point *P*, then *E* ⊥ *l* at *P*.

4) Two Planes

For any two planes, there are three possible cases :

① the planes intersect,

② the planes are parallel, or

③ the planes are congruent.

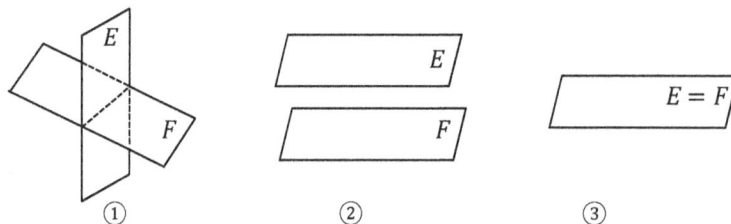

3. Perpendicular Lines and Planes

(1) The Perpendicular Bisector

For any given plane, the perpendicular bisector of a segment $\overline{AB}$ is the line which is perpendicular to the segment at its midpoint.

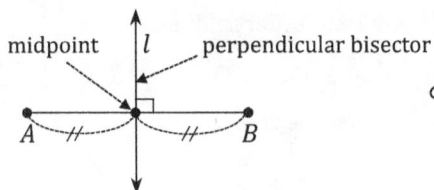

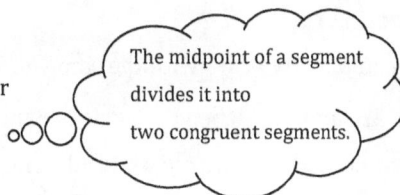

midpoint l perpendicular bisector

A B

The midpoint of a segment divides it into two congruent segments.

If a point P is on the perpendicular bisector l of the segment $\overline{AB}$, the distance from P to A equals the distance from P to B ($PA = PB$).

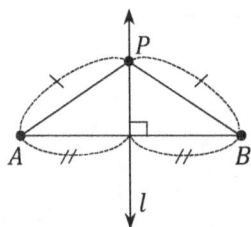

P

A B

l

Conversely, if $PA = PB$, then P is on the perpendicular bisector l.

(2) The Perpendicular Segment

The distance to a plane from a point which is not contained in the plane is the length of the perpendicular segment from the point to the plane.

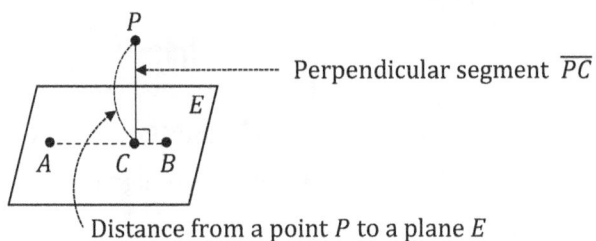

P

Perpendicular segment $\overline{PC}$

E

A C B

Distance from a point P to a plane E

1-4   Angular Measures

1. Angles

An angle is created by two rays, not by two segments.

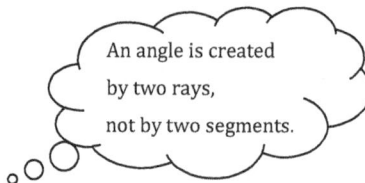

(1) Angle, Side, and Vertex

An *angle* is created by two rays with the same end point but different lines

If the rays are $\overrightarrow{AB}$ and $\overrightarrow{AC}$, then the angle is denoted by $\angle A$, $\angle BAC$, or $\angle CAB$.

The two rays are called its *sides* and the end point is called the *vertex*.

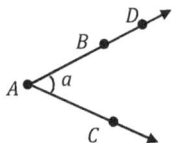

The sides of $\angle A$ are ($\overrightarrow{AB}$ and $\overrightarrow{AC}$) or ($\overrightarrow{AD}$ and $\overrightarrow{AC}$).

The vertex of $\angle A$ is a point A.

The angle A is $\angle a = \angle A = \angle BAC = \angle CAB = \angle DAC = \angle CAD$.

Note : The interior and exterior of an angle :

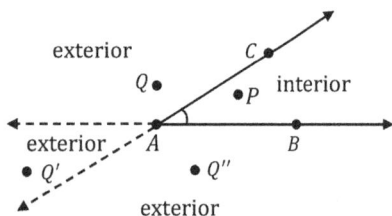

A point P is located in the interior of $\angle CAB$

if points C and P are on the same side of the line $\overleftrightarrow{AB}$ and

points P and B are on the same side of the line $\overleftrightarrow{AC}$.

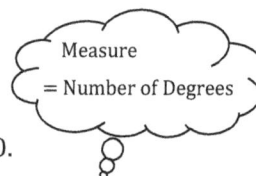

(2) Classifying Angles

Measure = Number of Degrees

For any angle, there is a corresponding real number between 0 and 180.

The measure of an angle A is denoted by $m \angle A$ and classified according to its measure.

1)   Acute Angles

An acute angle is an angle with a measure greater than 0° and less than 90°.

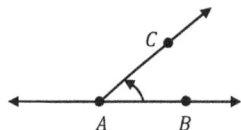

$$0° < m \angle A < 90°$$

2) Right Angles

A right angle is an angle with the measure 90°.

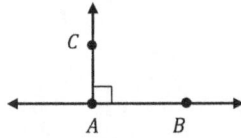

$$m \angle A = 90°$$

If $\overrightarrow{AB}$ and $\overrightarrow{AC}$ form a right angle, then $\overrightarrow{AB}$ and $\overrightarrow{AC}$ are called *perpendicular*, and denoted by $\overrightarrow{AB} \perp \overrightarrow{AC}$.

3) Obtuse Angles

An obtuse angle is an angle with a measure greater than 90° and less than 180°.

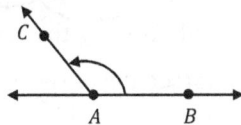

$$90° < m \angle A < 180°$$

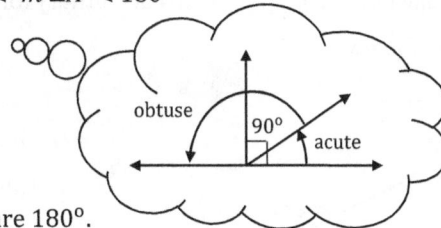

4) Straight Angles

A straight angle is an angle with the measure 180°.

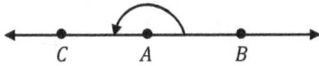

(3) Supplementary and Complementary

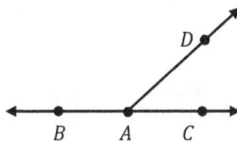

If $\overrightarrow{AB}$ and $\overrightarrow{AC}$ are opposite rays, and $\overrightarrow{AD}$ is any other ray, then $\angle BAD$ and $\angle DAC$ form a linear pair.

If the sum of the measures of two angles is 180°, then the angles are called *supplementary*.

Thus, $\angle DAC$ is a supplement of $\angle BAD$.

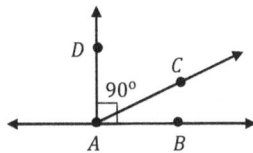

If the sum of the measures of two angles is 90°, then the angles are called *complementary*.

Thus, $\angle CAB$ is a complement of $\angle DAC$.

(4) Congruence

Two angles with the same measure are congruent.

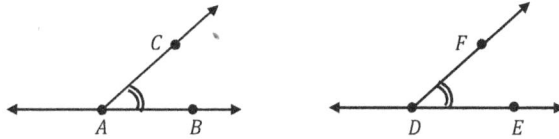

If $m\angle CAB = m\angle FDE$, then $\angle CAB$ and $\angle FDE$ are *congruent* and written $\angle CAB \cong \angle FDE$.

($\angle CAB$ is congruent to $\angle FDE$)

> Adjacent angles share a common side and a common vertex (corner point), but do not overlap.

(5) Vertical Angles

When two lines intersect at one point, four angles are formed. Among the four angles, the pairs of non-adjacent angles (opposite each other) formed by the intersection of two straight lines are called *vertical angles*.

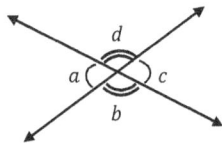

$\angle a$ and $\angle c$ are vertical angles,

and also $\angle b$ and $\angle d$ are vertical angles.

> If two lines intersect perpendicularly, four right angles are formed.

Note : Vertical angles are congruent.

2. Angles formed by a Transversal

> A transversal line crosses at least two other lines.

(1) Transversals

A transversal of two lines is a line which intersects them in two distinct points.

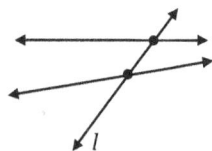

l is a transversal.

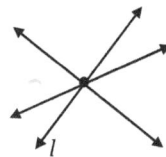

l is not a transversal.

(2) Corresponding Angles and Alternate Interior Angles

1) Corresponding angles are angles located at the same relative position on each line.

2) Alternate interior angles are the angles between the lines on alternate sides of the transversal.

3) The vertical angles of alternate interior angles are alternate exterior angles.

∠c and ∠e are alternate interior angles. So, the vertical angles of ∠c and ∠e are ∠a and ∠g, respectively. Therefore, ∠a and ∠g are alternate exterior angles.

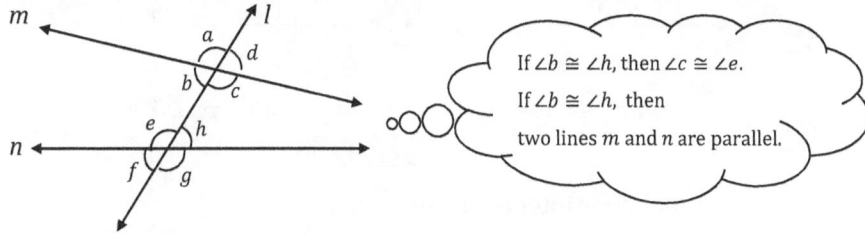

If ∠b ≅ ∠h, then ∠c ≅ ∠e.
If ∠b ≅ ∠h, then
two lines m and n are parallel.

corresponding angles	alternate interior angles	alternate exterior angles
∠a and ∠e	∠b and ∠h	∠a and ∠g
∠b and ∠f	∠c and ∠e	∠d and ∠f
∠c and ∠g		
∠d and ∠h		

4) If ∠a and ∠b are alternate interior angles and if ∠a and ∠c are vertical angles, then ∠c and ∠b are corresponding angles.

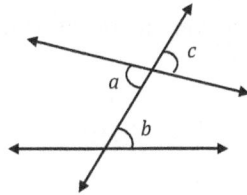

Note : If a pair of corresponding angles are congruent, then a pair of alternate interior angles are congruent.

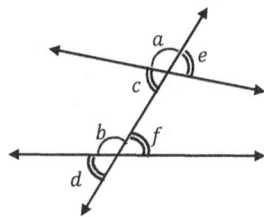

If ∠a ≅ ∠b (corresponding angles), then ∠c ≅ ∠d.
(∵ $m∠a + m∠C = 180°$ and $m∠b + m∠d = 180°$.
Since ∠a ≅ ∠b, ∠c ≅ ∠d.)

Since ∠e and ∠f are vertical angles of ∠c and ∠d, respectively, if ∠c ≅ ∠d, then ∠e ≅ ∠f.
Therefore, ∠c ≅ ∠f (alternate interior angles).

(3) Angles and Parallel Lines

1) For two lines intersected by a transversal,

    ① if a pair of corresponding angles are congruent, then the two lines are parallel.

    ② if a pair of alternate interior angles are congruent, then the two lines are parallel.

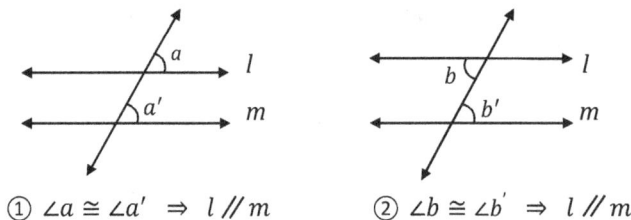

    ① $\angle a \cong \angle a' \Rightarrow l \mathbin{/\mkern-4mu/} m$ ② $\angle b \cong \angle b' \Rightarrow l \mathbin{/\mkern-4mu/} m$

2) For two parallel lines intersected by a transversal,

    ① each pair of corresponding angles are congruent.

    ② each pair of alternate interior angles are congruent.

    ③ each pair of alternate exterior angles are congruent

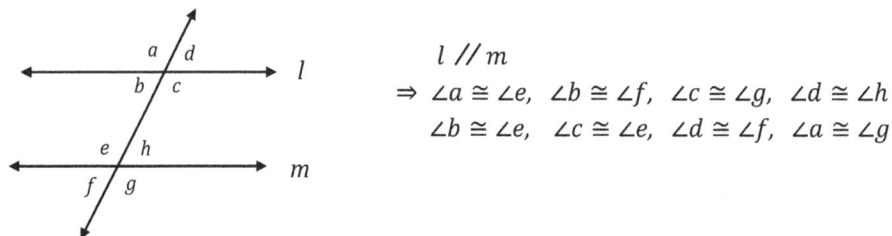

$l \mathbin{/\mkern-4mu/} m$
$\Rightarrow \angle a \cong \angle e, \ \angle b \cong \angle f, \ \angle c \cong \angle g, \ \angle d \cong \angle h$
$\quad \angle b \cong \angle e, \ \angle c \cong \angle e, \ \angle d \cong \angle f, \ \angle a \cong \angle g$

    ④ the interior angles on the same side of the transversal are supplementary.

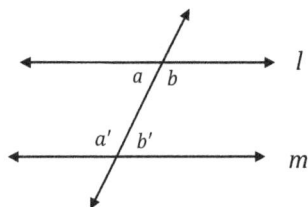

Note :   *If $l \mathbin{/\mkern-4mu/} m \Rightarrow \angle a \cong \angle b'$ (alternate interior angles)*

      Since $m\angle a + m\angle b = 180°$ and $m\angle a' + m\angle b' = 180°$, $\angle b \cong \angle a'$

      Then, $m\angle a + m\angle a' = 180°$ and $m\angle b + m\angle b' = 180°$.

      That is, $\angle a$ is supplementary of $\angle a'$ and $\angle b$ is supplementary of $\angle b'$.

Exercises

#1 Find the distance for the following from the figure below :

(1) The distance from point A to point B.

(2) The distance from point B to point D.

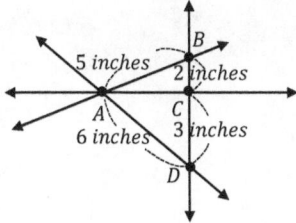

#2 Let points M and N be the midpoints of segments $\overline{AB}$ and $\overline{MB}$, respectively.

(1) Determine whether the following expressions are the true or false.

1) $AM = BM$

2) $AM = \frac{1}{2}AB$

3) $NB = \frac{1}{3}AB$

4) $AB = 2MN$

5) $MN = \frac{1}{2}AM$

(2) If the distance AB is 20 inches, find the distance MN and the distance AN.

#3 Determine whether each statement is true or false.

(1) If two lines are parallel to another line, then the two lines are parallel to each other.

(2) Two lines on a plane always intersect at one point.

(3) If two lines are not intersecting, then the lines are not on a plane.

(4) There is always a plane containing two distinct lines.

(5) Two parallel lines do not have intersection points.

(6) If two distinct lines intersect, their intersection contains only one point.

(7) If two points of a line lie in a plane, then the line lies in the same plane.

(8) If two different planes intersect, their intersection is a point.

#4 Using the figure, find the measure of each angle.

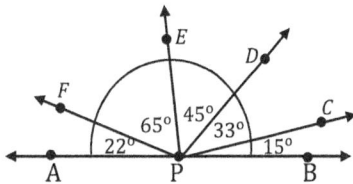

(1) $\angle BPC$

(2) $\angle CPE$

(3) $\angle CPF$

(4) $\angle APE$

#5 Using the figure, identify each angle as acute, right, or obtuse.

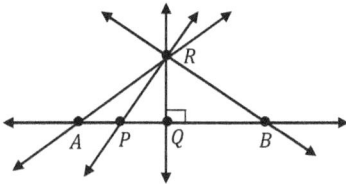

(1) $\angle RAB$

(2) $\angle PRQ$

(3) $\angle RQP$

(4) $\angle APR$

#6 Determine the measure of the complement of an angle with a measure of :

(1) $36°$

(2) $75°$

(3) $24.5°$

(4) $45° + n°$

(5) $90° - n°$

#7 Using the figure, answer the following :

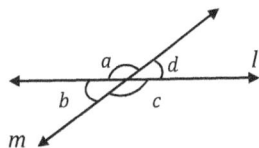

(1) If $m \angle a = 120°$, what is $m \angle b$?

(2) If $m \angle d = 43°$, what are $m \angle a$, $m \angle b$, and $m \angle c$?

#8 For the following figures, find the measures of $\angle a$:

(1)

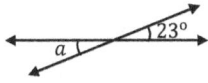

(2)

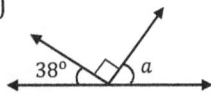

(3)

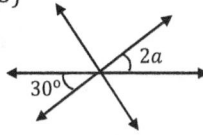

(4)

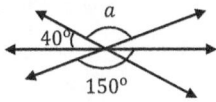

(5)

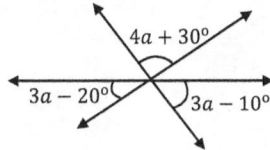

#9 Lines l and m are parallel. Using the following figures, find the measures of the angles.

(1)

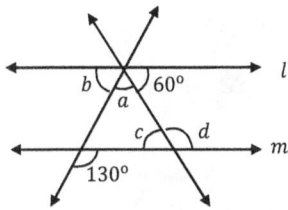

① $\angle a$ ② $\angle b$ ③ $\angle c$ ④ $\angle d$

(2)

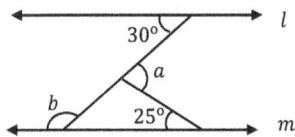

① $\angle a$ ② $\angle b$

Triangles and Polygons

Chapter 2   Triangles and Polygons

CHAPTER
2

Chapter 2. Triangles and Polygons

2-1 Triangles

1. Defining a Triangle

For any three non-collinear points A, B, and C,

the connection of the three segments $\overline{AB}, \overline{BC}$, and $\overline{CA}$ is called a *triangle*, and is denoted by $\triangle ABC$.

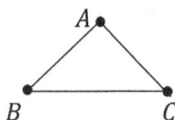

The segments $\overline{AB}, \overline{BC}$, and $\overline{CA}$ are called its *sides*

and the points A, B, and C at which two sides intersect are called

its *vertices*.

angle opposite the side $\overline{BC}$

angle opposite the side $\overline{AB}$

Note : The interior and exterior of a triangle :

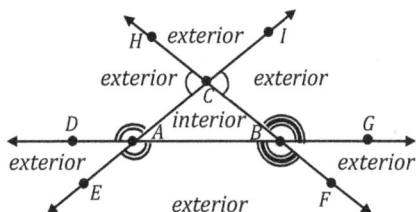

angle opposite the side $\overline{AC}$

The sides opposite the angles $\angle A, \angle B$, and $\angle C$

are $\overline{BC}, \overline{AC}$, and $\overline{AB}$, respectively.

The interior of a triangle includes all the segments on the triangle and all the points lying in the interior of each angle of the triangle. The interior angles of a $\triangle ABC$ are $\angle CAB, \angle ABC$, and $\angle ACB$.

There are six exterior angles of $\triangle ABC$:

$\angle CAD, \angle BAE$ *(supplements of $\angle A$),* $\angle ABF, \angle CBG$ *(supplements of $\angle B$), $\angle HCA$,*

$\angle ICB$ *(supplements of $\angle C$)*

The interior angles are located inside
the triangle and the exterior angles are
supplements of the interior angles.

To form a triangle, one of the following cases must be there :

Case 1. When the lengths of three sides are given, a triangle can be formed.

(The length of the longest side must be less than the sum of the lengths of the two remaining

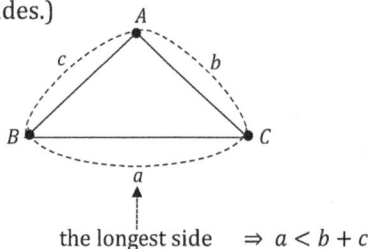

sides.)

the longest side $\Rightarrow a < b + c$

Case 2. When the lengths of two sides and the included angle are given, a triangle can be formed.

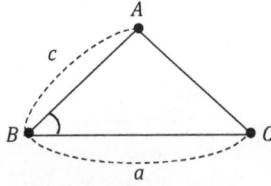

Case 3. When two angles and the length of the included side are given, a triangle can be formed.

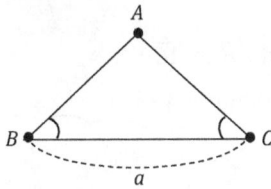

2. Classifying Triangles

(1) Isosceles

A triangle with at least two congruent sides is called *isosceles*.

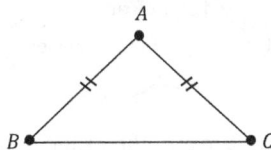

1) Leg, Base, Base Angle, and Vertex Angle

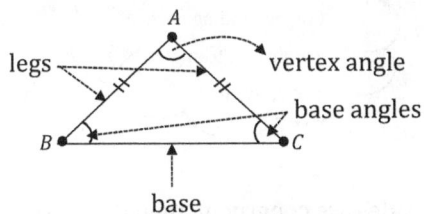

Exactly two sides of an isosceles triangle are congruent. The congruent sides are the *legs* of an isosceles triangle and the remaining side is the *base*. The two angles including the base are the *base angles* and the angle opposite the base is the *vertex angle*.

(2) Equilateral

A triangle with three congruent sides is called *equilateral*.

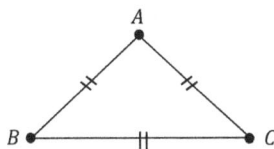

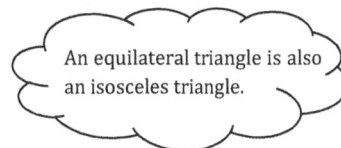

An equilateral triangle is also an isosceles triangle.

(3) Equiangular

A triangle with three congruent angles is called *equiangular*.

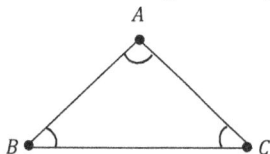

(4) Scalene

A triangle with no congruent sides is called *scalene*.

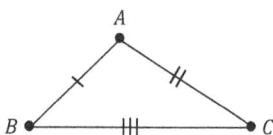

$\overline{AB} \ncong \overline{BC}$, $\overline{BC} \ncong \overline{CA}$, and $\overline{AB} \ncong \overline{AC}$

(three sides of different lengths)

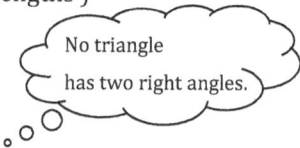

No triangle has two right angles.

(5) Right

A triangle which includes a right angle is called *right*.

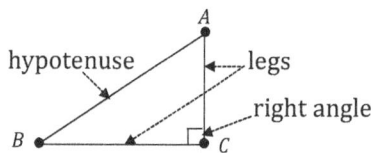

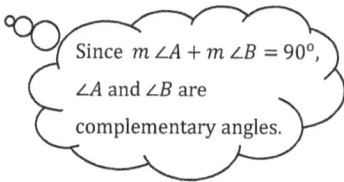

Since $m\angle A + m\angle B = 90°$, $\angle A$ and $\angle B$ are complementary angles.

The longest side opposite the right angle is called the *hypotenuse* and the other two sides are called the *legs*.

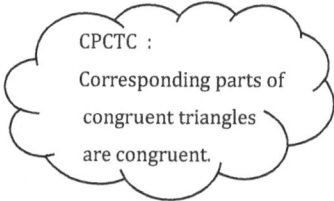

CPCTC : Corresponding parts of congruent triangles are congruent.

3. Congruence of Triangles

(1) SSS Correspondence and Postulate

For a pair of triangles, if all three corresponding sides are congruent, then the correspondence is called a *SSS correspondence*.

If three sides of one triangle are congruent to three sides of another triangle, then the two triangles are congruent. *(Side-Side-Side Postulate)*

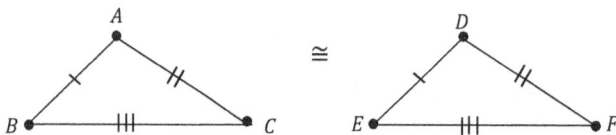

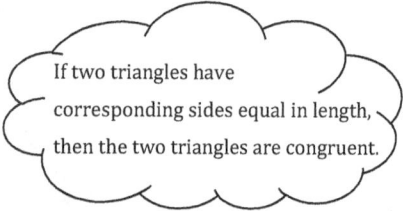

If two triangles have corresponding sides equal in length, then the two triangles are congruent.

If $\overline{AB} \cong \overline{DE}$, $\overline{BC} \cong \overline{EF}$, and $\overline{AC} \cong \overline{DF}$, then $\triangle ABC \cong \triangle DEF$.

(2) SAS Correspondence and Postulate

For a pair of triangles, if two pairs of corresponding sides are congruent, and a pair of corresponding included angles is congruent, then the correspondence is called a *SAS correspondence.*

If two sides and the included angle of one triangle are congruent to the corresponding parts of another triangle, then the two triangles are congruent. *(Side-Angle-Side Postulate)*

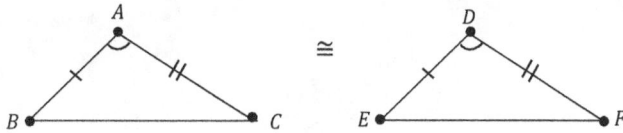

If $\overline{AB} \cong \overline{DE}$, $\overline{AC} \cong \overline{DF}$, and $\angle A \cong \angle D$, then $\triangle ABC \cong \triangle DEF$.

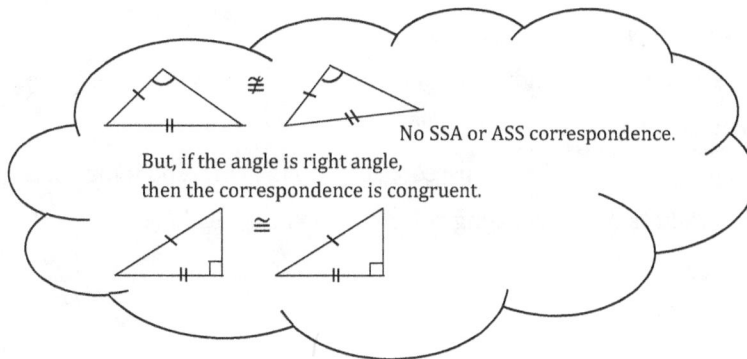

No SSA or ASS correspondence.
But, if the angle is right angle,
then the correspondence is congruent.

(3) ASA Correspondence and Postulate

For a pair of triangles, if two pairs of corresponding angles and a pair of corresponding included sides are congruent, then the correspondence is called an *ASA correspondence.*

If two angles and the included side of one triangle are congruent to the corresponding parts of another triangle, then the two triangles are congruent. *(Angle-Side-Angle Postulate)*

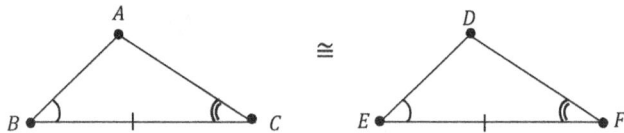

If $\angle B \cong \angle E$, $\angle C \cong \angle F$, and $\overline{BC} \cong \overline{EF}$, then $\triangle ABC \cong \triangle DEF$.

(4) SAA (or AAS) Correspondence and Postulate

For a pair of triangles, if a pair of corresponding sides is congruent and two pairs of corresponding angles are congruent, then the correspondence is called a *SAA correspondence* or *AAS correspondence.* If two angles and the non-included side of one triangle are congruent to the corresponding parts of another triangle, then the two triangles are congruent. *(Side-Angle-Angle Postulate or Angle-Angle-Side Postulate)*

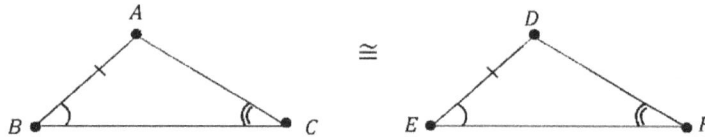

If $\angle B \cong \angle E$, $\angle C \cong \angle F$, and $\overline{AB} \cong \overline{DE}$, then $\triangle ABC \cong \triangle DEF$.

(5) HL Correspondence and Theorem

For a pair of right triangles, if the corresponding hypotenuses and corresponding legs are congruent, then the correspondence is called a *HL correspondence.*

If the hypotenuse and a leg of one right triangle are congruent to the corresponding parts of another right triangle, then the two right triangles are congruent. *(Hypotenuse-Leg Theorem)*

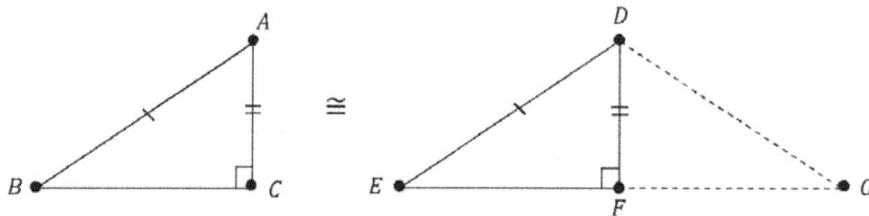

Let $m \angle C = m \angle F = 90°$, $\overline{AB} \cong \overline{DE}$ and $\overline{AC} \cong \overline{DF}$.

If a point G satisfies $\overline{BC} \cong \overline{FG}$ (equal distance), we get $\triangle DEF \cong \triangle DFG$.

($\because$ Since $\overline{AB} \cong \overline{DG}$, $\angle B \cong \angle G$.

Since $\overline{DE} \cong \overline{DG}$, $\angle E \cong \angle G$.

So, $\triangle DEF \cong \triangle DFG$.)

Therefore, $\triangle ABC \cong \triangle DEF$

2-2 Isosceles triangles

1. Bisectors of Angles

If a point P is in the interior of $\angle BAC$ and $\angle BAP \cong \angle PAC$, then $\overrightarrow{AP}$ bisects $\angle A$ and $\overrightarrow{AP}$ is called the *bisector* of $\angle A$.

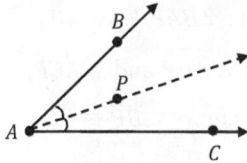

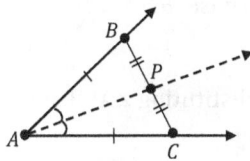

If $\overline{AB} \cong \overline{AC}$ and an interior point P is the midpoint of $\overline{BC}$,

Then $\triangle ABP \cong \triangle ACP$ by SSS Postulate.

So, $\angle BAP \cong \angle PAC$.

Therefore, $\angle A$ has a bisector $\overrightarrow{AP}$.

Note : Every angle has only one bisector.

2. Properties of Isosceles Triangles

(1) If a triangle $\triangle ABC$ is isosceles, then base angles are congruent.

That is, $\overline{AB} \cong \overline{AC} \Rightarrow \angle B \cong \angle C$

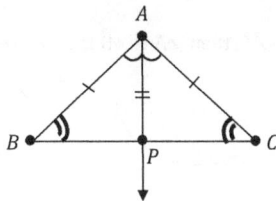

($\because$ Since $\triangle ABC$ is isosceles,

two sides of $\triangle ABC$ are congruent ($\overline{AB} \cong \overline{AC}$).

Since $\angle A$ has the angle bisector $\overrightarrow{AP}$, $\angle BAP \cong \angle CAP$.

By SAS Postulate, $\triangle ABP \cong \triangle APC$.

Therefore, $\angle B \cong \angle C$ (The base angles are congruent.))

Note : If two sides of a triangle are congruent, then the base angles are also congruent.

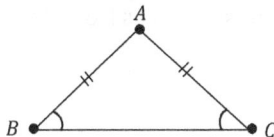

If $\overline{AB} \cong \overline{AC}$, then $\angle B \cong \angle C$.

Conversely,

if $\angle B \cong \angle C$, then $\overline{AB} \cong \overline{AC}$.

Base angles of an isosceles triangle are congruent and have the same measure.

(2) If $\triangle ABC$ is isosceles and $\overrightarrow{AP}$ is a bisector of the vertex angle, then $\overrightarrow{AP}$ is the perpendicular bisector of the base.

That is, $\overline{AB} \cong \overline{AC}$, and $\angle BAP \cong \angle CAP$ $\Rightarrow$ $\overline{AP} \perp \overline{BC}$ and $\overline{BP} \cong \overline{PC}$

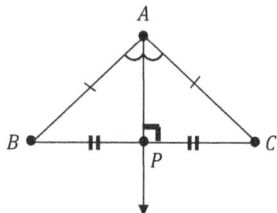

($\because$ Since $\triangle ABC$ is isosceles,

two sides of $\triangle ABC$ are congruent ($\overline{AB} \cong \overline{AC}$).

Since $\overrightarrow{AP}$ is the bisector of $\angle A$, $\angle BAP \cong \angle CAP$.

Since $\overrightarrow{AP}$ is a common side of $\triangle ABP$ and $\triangle ACP$,

$\triangle ABP \cong \triangle ACP$ by SAS Postulate. $\therefore$ $\overline{BP} \cong \overline{PC}$

That means the point P is the midpoint of $\overline{BC}$. $\therefore$ $\overrightarrow{AP}$ bisects the base $\overline{BC}$.

Since $\triangle ABP \cong \triangle ACP$, $\angle APB \cong \angle APC$.

Since $m \angle APB + m \angle APC = 180°$, $m \angle APB + m \angle APB = 180°$ substituting $\angle APB$ into $\angle APC$.

That is, $2(m \angle APB) = 180°$. $\therefore$ $m \angle APB = 90°$. So, $\overline{AP} \perp \overline{BC}$.

Therefore, $\overrightarrow{AP}$ is the perpendicular bisector of the base.)

2-3 Equilateral and Equiangular Triangles

1. Equilateral Triangles

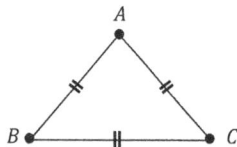

If $\angle A \cong \angle B \cong \angle C$, then $\overline{AB} \cong \overline{BC} \cong \overline{AC}$.
Conversely,
If $\overline{AB} \cong \overline{BC} \cong \overline{AC}$, then $\angle A \cong \angle B \cong \angle C$.

If a triangle is equilateral, all three of its sides are congruent.

First, consider $\overline{AB} \cong \overline{AC}$. Then the triangle is isosceles. So, $\angle B \cong \angle C$.

Next, consider $\overline{AB} \cong \overline{BC}$. In this case, the triangle can still be considered as isosceles. So, $\angle A \cong \angle C$.

Thus, $\angle A \cong \angle B \cong \angle C$ (all three angles are congruent).

Therefore, every equilateral triangle is both isosceles and equiangular.

2. Equiangular Triangles

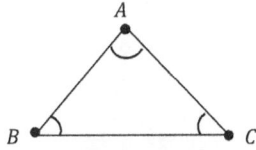

If a triangular is equiangular, then all three of its angles are congruent.

At first, consider $\angle B \cong \angle C$. Then, $\overline{AB} \cong \overline{AC}$ by the properties of isosceles triangles.

Now, consider $\angle A \cong \angle C$. Then, $\overline{AB} \cong \overline{BC}$ by the properties of isosceles triangles.

So, $\overline{AB} \cong \overline{BC} \cong \overline{AC}$ (all three sides are congruent).

Therefore, every equiangular triangle is both isosceles and equilateral.

2-4 Right Triangles

1. Congruence

(1) RHS (Right-Hypotenuse-Side) Correspondence = HL theorem

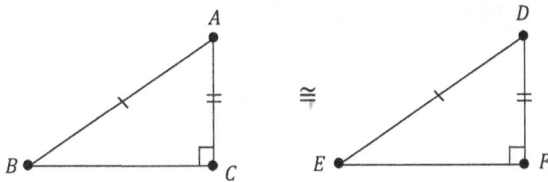

For a pair of right triangles, if the corresponding hypotenuses and corresponding legs are congruent, then the correspondence is congruent.

(2) RHA (Right-Hypotenuse-Angle) Correspondence

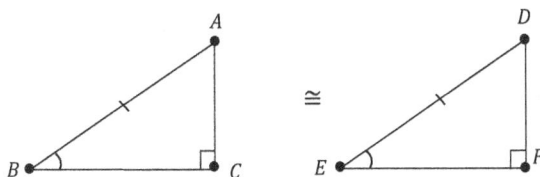

For a pair of right triangles, if the corresponding hypotenuses and corresponding acute angles are congruent, then the correspondence is congruent.

($\because$ Since $m \angle C = m \angle F = 90°$, $\angle C \cong \angle F$.

Since $\overline{AB} \cong \overline{DE}$ and $\angle B \cong \angle E$, $\triangle ABC \cong \triangle DEF$ by SAA Postulate.)

2. Bisectors of angles

Let $\triangle ABP$ and $\triangle ACP$ be right angles with $m\angle B = m\angle C = 90°$.

(1) If $\overrightarrow{AP}$ is a bisector of $\angle A$, then $\overline{PB} \cong \overline{PC}$.

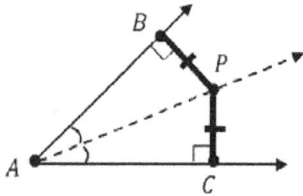

Since $\angle B \cong \angle C$, $\angle BAP \cong \angle CAP$,

and $\overline{AP}$ is a common side of $\triangle ABP$ and $\triangle ACP$,

$\triangle ABP \cong \triangle ACP$

by RHA correspondence (or SAA Postulate).

Therefore, $\overline{PB} \cong \overline{PC}$.

(2) If $\overline{PB} \cong \overline{PC}$, then $\overrightarrow{AP}$ is a bisector of $\angle A$.

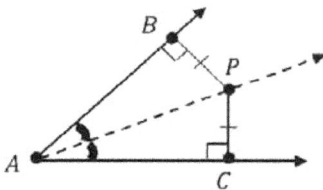

Since $\angle B \cong \angle C$, $\overline{PB} \cong \overline{PC}$,

and $\overline{AP}$ is a common side of $\triangle ABP$ and $\triangle ACP$,

$\triangle ABP \cong \triangle ACP$

by RHS correspondence (HL Theorem).

$\therefore \ \angle BAP \cong \angle CAP$

Therefore, $\overrightarrow{AP}$ is a bisector of $\angle A$.

3. The Midpoint of the Hypotenuse

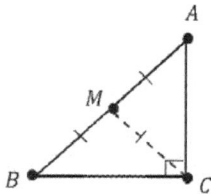

For a right triangle $\triangle ABC$ with $m\angle C = 90°$,

if M is a midpoint of the hypotenuse $\overline{AB}$,

then $AM = BM = CM$

($\because$ Consider a parallelogram $\square ADBC$.

Since M is a midpoint of $\overline{AB}$, $AB = AM + MB$.

Also, $DC = DM + MC$.

Since $AM = MB$ and $DM = MC$,

$AB = AM + MB = AM + AM = 2AM$ and

$DC = DM + MC = MC + MC = 2MC$.

Since $AB = DC$, $2AM = 2MC$; $AM = MC$.

Therefore, $AM = BM = CM$.)

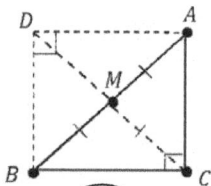

The diagonals of a parallelogram bisect each other.
⇒ See Chapter 3, 3-3 for more information!

4. Similarity

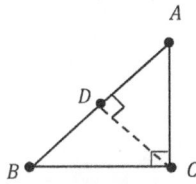

For a right triangle $\triangle ABC$ with $m \angle C = 90^\circ$,

if $\overline{CD} \perp \overline{AB}$, then $\triangle ABC \sim \triangle ACD$ and $\triangle ABC \sim \triangle CBD$.

Therefore, $\triangle ACD \sim \triangle CBD$.

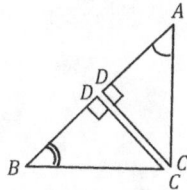
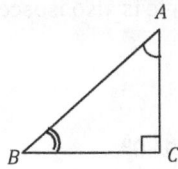

(∵ Consider these three right triangles.

By AA Similarity Theorem,

$\triangle ABC \sim \triangle ACD$ and $\triangle ABC \sim \triangle CBD$)

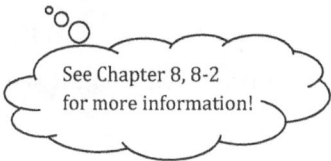

See Chapter 8, 8-2
for more information!

Note : AA (Angle-Angle) Similarity Theorem

For two corresponding triangles,

if two pairs of corresponding angles are congruent, then the two triangles are similar.

Example

Let $AD = 2$ and $BD = 8$. Find CD.

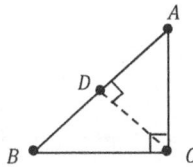

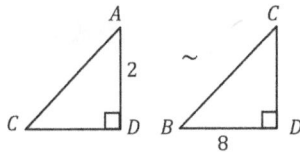

To create similar right triangles,
consider the altitude drawn to the hypotenuse
of the right triangle.
(Altitude is a line that extends from one vertex
of a triangle perpendicular to the opposite side.)

Note that $\overline{AC}$ is the hypotenuse of $\triangle ACD$ and $\overline{CB}$ is the hypotenuse of $\triangle CBD$.

If $AD = 2$ and $BD = 8$, then $CD : 2 = 8 : CD$

∴ $CD^2 = 16$.

Therefore, $CD = 4$

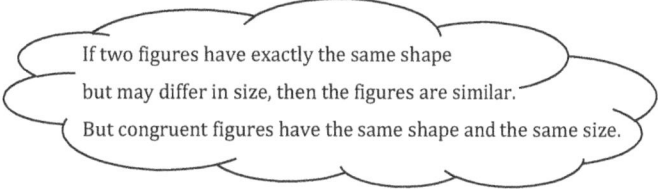

If two figures have exactly the same shape
but may differ in size, then the figures are similar.
But congruent figures have the same shape and the same size.

5. Special right Triangles

(1) $45° - 45° - 90°$ Triangle

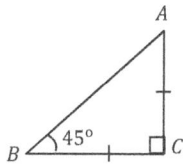

For a right triangle $\triangle ABC$ with $m \angle C = 90°$,

if one of the acute angles is $45°$,

then the right triangle is also isosceles.

($\because$ Since the sum of the interior angles of a triangle is $180°$,

$m \angle A + 45° + 90° = 180°$. $\therefore m \angle A = 45°$

Since $m \angle A = m \angle B = 45°$, $\angle A \cong \angle B$.

So, $\overline{AC} \cong \overline{BC}$

Therefore, $\triangle ABC$ is isosceles.)

(2) $30° - 60° - 90°$ Triangle

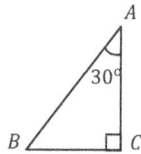

For a right triangle $\triangle ABC$ with $m \angle C = 90°$,

if one of the acute angles is $30°$,

then the other acute angle is $60°$.

Let $m \angle A = 30°$. Then $m \angle B = 60°$.

Since $\angle A$ is the smallest angle of $\triangle ABC$, $\overline{BC}$ is the shortest side opposite the smallest angle $\angle A$.

Consider a right triangle $\triangle ACD$ with $m \angle C = 90°$.

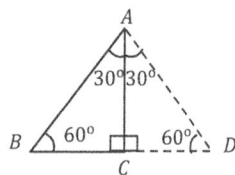

If $m \angle CAD = 30°$,

$\overline{AC}$ is the bisector of $\angle BAD$.

$\therefore \overline{BC} \cong \overline{CD}$

Since $\triangle ABD$ is a regular triangle, $BC = \frac{1}{2} AB$.

Therefore, the length of the shortest side in a $30° - 60° - 90°$ triangle is half the length of the hypotenuse.

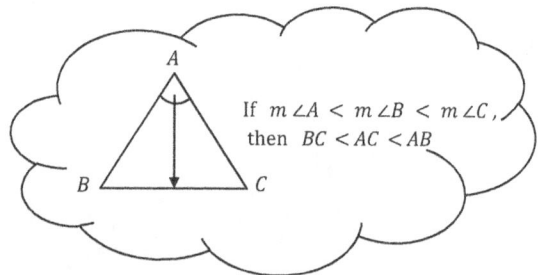

If $m \angle A < m \angle B < m \angle C$, then $BC < AC < AB$

2-5 Polygons

1. Classifying Polygons

(1) Definition

> A collinear set is a set of points lying on a single line.

A polygon is the connection of n segments, where $n \geq 3$, such that

1) no two of the segments intersect except at their end points and

2) no two segments with common end points are collinear.

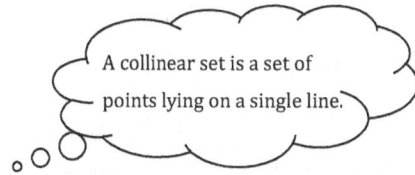

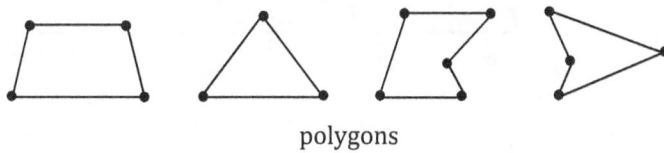

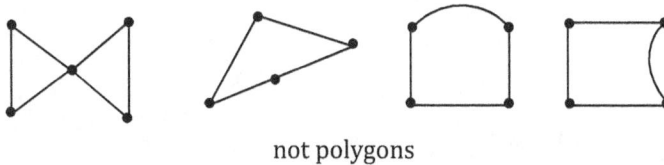

polygons

not polygons

The segments are called the *sides* of the polygon and the end points of the segments are called the *vertices* of the polygon.

The sum of the length of the sides is called the *perimeter* of the polygon.

A polygon with n sides ($n \geq 3$) is called a *n-gon*.

A diagonal of a polygon is a segment connecting two non-adjacent vertices.

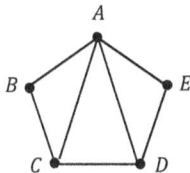

Diagonals from a vertex A

: $\overline{AC}$, $\overline{AD}$

Diagonals divide the interior of the polygon (n-gon) into $(n - 2)$ non-overlapping triangles.

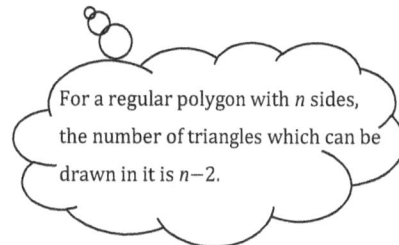

> For a regular polygon with n sides, the number of triangles which can be drawn in it is $n-2$.

Number of sides (n)	name of polygon (n-gon)	number of diagonals from a vertex: ($n-3$)	total number of diagonals: $\frac{n(n-3)}{2}$
3	triangle (3-gon)	0	0
4	quadrilateral (4-gon)	1	2
5	pentagon (5-gon)	2	5
6	hexagon (6-gon)	3	9
7	heptagon (7-gon)	4	14
8	octagon (8-gon)	5	20
9	nonagon (9-gon)	6	27
10	decagon (10-gon)	7	35

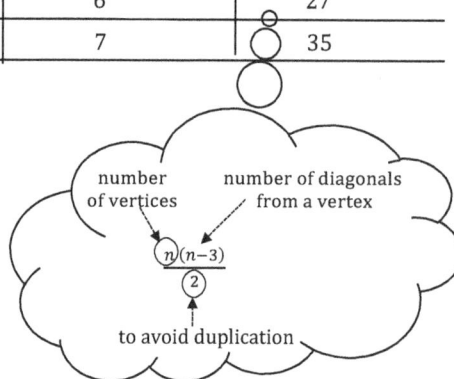

number of vertices

number of diagonals from a vertex

$$\frac{n(n-3)}{2}$$

to avoid duplication

Note : ① *If no extending lines of the sides of a polygon intersect the interior of the polygon,*

then the polygon is called convex.

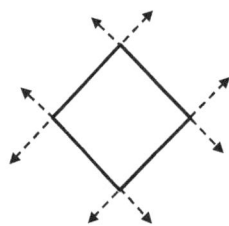

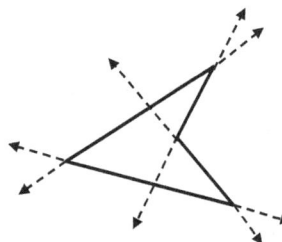

convex   not convex

② *If a polygon is convex, all of its sides are congruent, and all of its angles are congruent,*

then the polygon is called a regular polygon.

Square : regular polygon   trapezoid : not a regular polygon

2. Measures of the Angles of Convex Polygons

(1) Measures of Interior Angles

The sum of the measures of the interior angles of a triangle is $180°$.

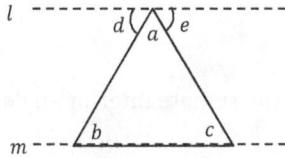

Consider $l \mathbin{/\mkern-5mu/} m$.

Since $\angle d$ and $\angle b$ are alternate interior angles, $\angle d \cong \angle b$.

Similarly, $\angle e \cong \angle c$.

Since $m\angle d + m\angle a + m\angle e = 180°$,

$m\angle b + m\angle a + m\angle c = 180°$.

Therefore, $m\angle a + m\angle b + m\angle c = 180°$.

Since diagonals from a vertex of a convex polygon (n-gon) divide the interior of the polygon into $(n-2)$ non-overlapping triangles, the sum of the measures of the interior angles of the n-sided convex polygon is $(n-2) \times 180°$.

Therefore, the measure of each interior angle for a regular n-sided polygon is $\frac{(n-2) \times 180°}{n}$.

Example

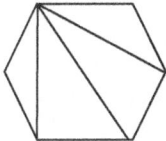

Imagine a regular hexagon.

The diagonals from a vertex divide the hexagon into 4 triangles.

So, the sum of the interior angle measures is

$(6-2) \times 180° = 720°$.

Therefore, the measure of one interior angle of a regular hexagon is

$\frac{(n-2) \times 180°}{n} = \frac{(6-2) \times 180°}{6} = 120°$.

(2)   Measures of Exterior Angles

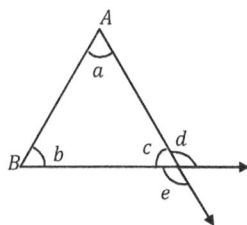

Every exterior angles of a triangle is supplements of one of the interior angles. The other two interior angles of the triangle are called the *remote interior angles*.

For example, $\angle a$ and $\angle b$ are the remote interior angles of the exterior angles $\angle d$ and $\angle e$.

The measure of an exterior angle is the sum of the measures of the remote interior angles.

That is, $m \angle d = m \angle a + m \angle b$.

($\because$ For a triangle $\triangle ABC$, the exterior angle $\angle d$ is a supplement of $\angle c$.

So, $m \angle d = 180° - m \angle c$.

Since $m \angle a + m \angle b + m \angle c = 180°$,

$m \angle d = 180° - m \angle c = (m \angle a + m \angle b + m \angle c) - m \angle c = m \angle a + m \angle b$.)

Example

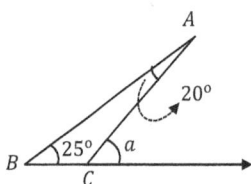

$m \angle a = 20° + 25° = 45°$

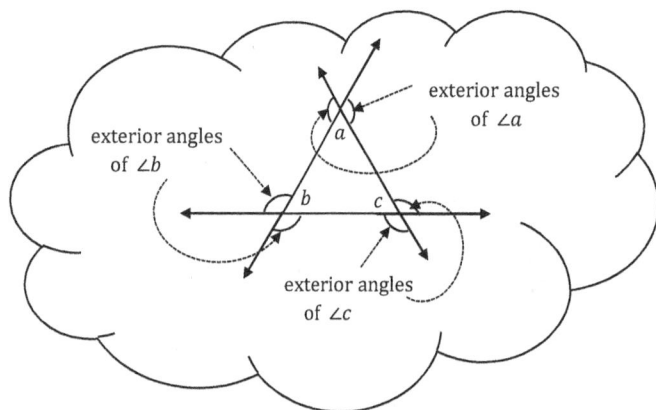

The sum of the measures of the interior and exterior angles in a regular polygon of n sides :

$n \times 180°$

—

The sum of the measures of the interior angles in a regular polygon of n sides :

$(n - 2) \times 180°$

=

The sum of the measures of the exterior angles in a regular polygon of n sides :

$n \times 180° - (n - 2) \times 180° = 360°$

Note 1 :

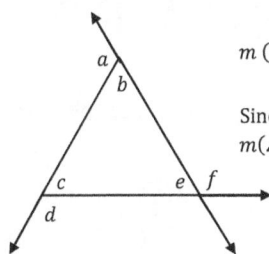

$m\left(\,(\angle a + \angle b) + (\angle c + \angle d) + (\angle e + \angle f)\,\right) - m(\,\angle b + \angle c + \angle e\,) = m(\angle a + \angle d + \angle f)$

Since $m\left(\,(\angle a + \angle b) + (\angle c + \angle d) + (\angle e + \angle f)\,\right) = 3 \times 180°$ and $m(\,\angle b + \angle c + \angle e\,) = 180°$,
$m(\angle a + \angle d + \angle f) = 360°$

Note 2 :

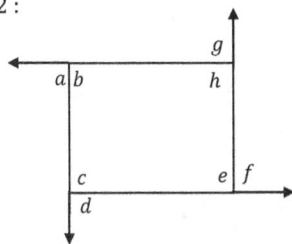

$m\left(\,(\angle a + \angle b) + (\angle c + \angle d) + (\angle e + \angle f) + (\angle g + \angle h)\,\right) - m(\,\angle b + \angle c + \angle e + \angle h\,)$

$= m(\angle a + \angle d + \angle f + \angle g)$

Since $m\left(\,(\angle a + \angle b) + (\angle c + \angle d) + (\angle e + \angle f) + (\angle g + \angle h)\,\right) = 4 \times 180°$

and $m(\,\angle b + \angle c + \angle e + \angle h\,) = (4 - 2) \times 180°$,

$m(\angle a + \angle d + \angle f + \angle g) = 360°$

Note 3 :

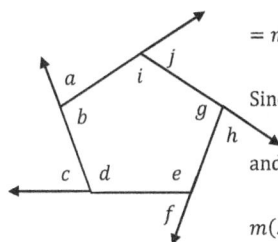

$m\left(\,(\angle a + \angle b) + (\angle c + \angle d) + (\angle e + \angle f) + (\angle g + \angle h) + (\angle i + \angle j)\,\right) - m(\,\angle b + \angle d + \angle e + \angle g + \angle i\,)$

$= m(\angle a + \angle c + \angle f + \angle h + \angle j)$

Since $m\left(\,(\angle a + \angle b) + (\angle c + \angle d) + (\angle e + \angle f) + (\angle g + \angle h) + (\angle i + \angle j)\,\right) = 5 \times 180°$

and $m(\,\angle b + \angle d + \angle e + \angle g + \angle i\,) = (5 - 2) \times 180°$,

$m(\angle a + \angle c + \angle f + \angle h + \angle j) = 360°$

The sum of the measures of the exterior angles of all convex polygons is 360°.

Thus , the sum of the measures of the exterior angles in a regular polygon of n sides is always $360°$*, regardless of the number*

of sides. Therefore, the measure of each exterior angle in a regular polygon of n sides is $\dfrac{360°}{n}$ *.*

Exercises

#1 Given the following lengths of sides, determine whether or not each example can be a triangle.

    (1) 4, 5, 6
    (2) 3, 4, 8
    (3) 5, 6, 11
    (4) 7, 7, 7

#2 Given the following side lengths, what kinds of triangles are each of the following?

    (1) 6, 8, 6
    (2) 5, 5, 5
    (3) 4, 5, 6
    (4) 3, 5, 7

#3 Identify which correspondence (SSS, SAS, ASA, HL, SAA, or none) must be applied for the following congruent triangles :

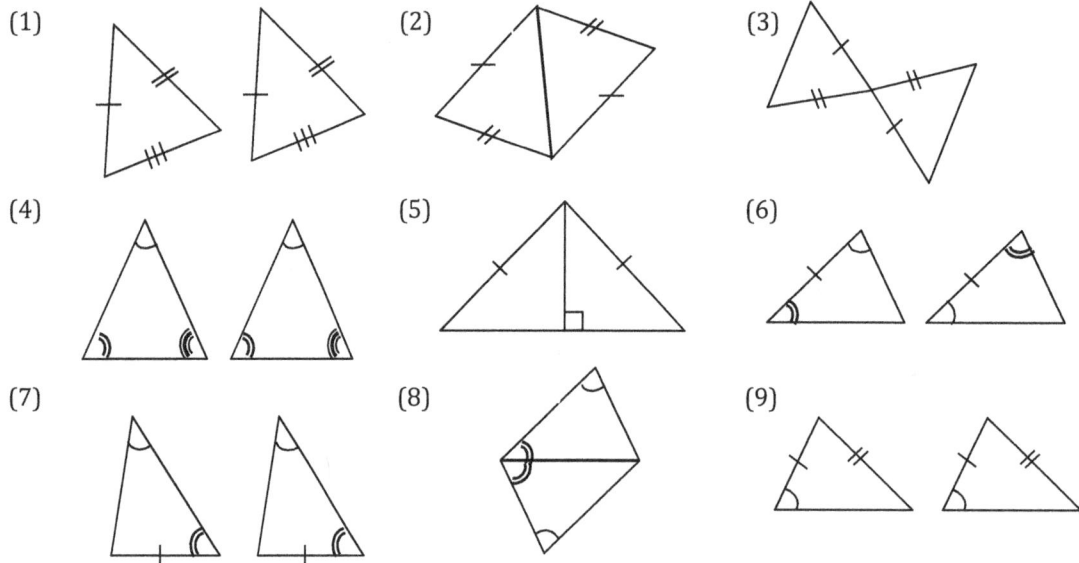

(1)　　　　　　(2)　　　　　　(3)

(4)　　　　　　(5)　　　　　　(6)

(7)　　　　　　(8)　　　　　　(9)

#4 Find the measure of angle $\angle a$ for the following :

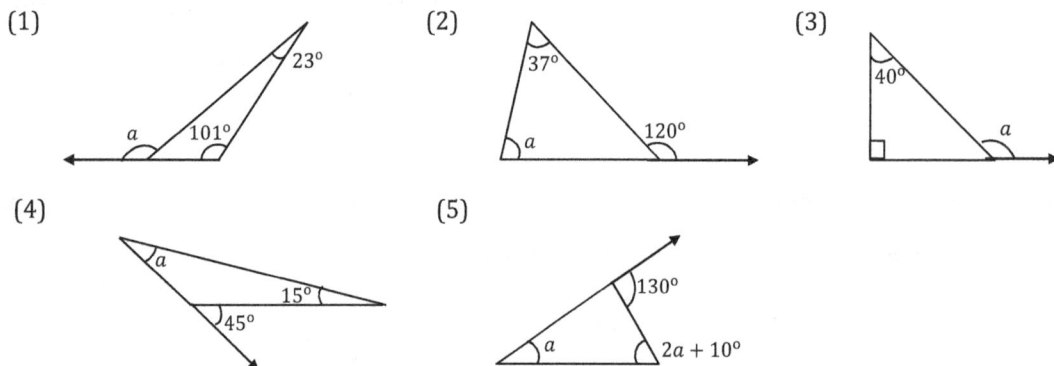

(1)　　　　　　(2)　　　　　　(3)

(4)　　　　　　(5)

(6)

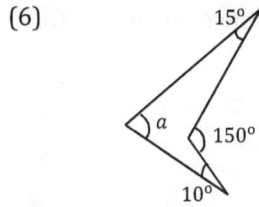

(7)

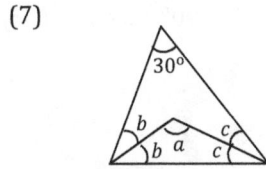

#5 Find the measure of $\angle a$.

(1)

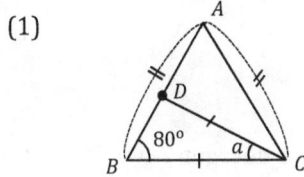

$\overline{AB} \cong \overline{AC}$, $\overline{BC} \cong \overline{CD}$, and
$m \angle B = 80^\circ$

(2)

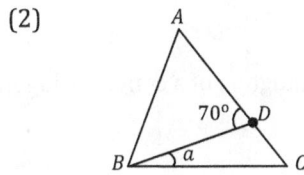

$\overline{AB} \cong \overline{AC} \cong \overline{BD}$ and
$m \angle ADB = 70^\circ$

(3)

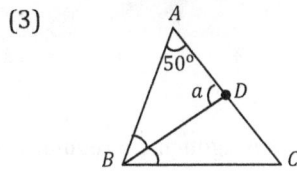

$\overline{AB} \cong \overline{AC}$, $m \angle A = 50^\circ$
and $\angle ABD \cong \angle DBC$

(4)

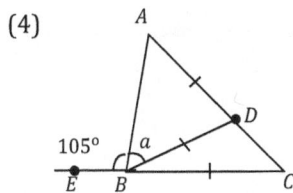

$\overline{AD} \cong \overline{BD} \cong \overline{BC}$ and
$m \angle ABE = 105^\circ$

(5)

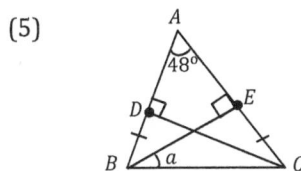

$m \angle A = 48^\circ$ and
$\overline{DB} \cong \overline{EC}$

(6)

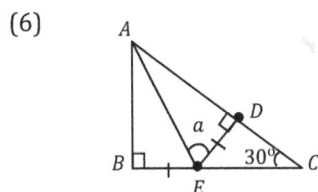

$m \angle B = 90^\circ$, $m \angle C = 30^\circ$,
$\overline{AC} \perp \overline{DE}$, $\overline{BE} \cong \overline{DE}$

#6 $\triangle ABC$ is a right triangle with $m \angle C = 90°$. Let $m \angle B = 30°$ and M be the midpoint of $\overline{AB}$.

Find $m \angle ACM$ and $m \angle BMC$.

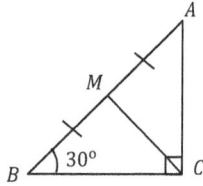

#7 How many diagonals has a polygon with 6 sides ; 12 sides ?

#8 How many triangles can be formed using the vertices of a hexagon ; decagon ?

#9 Find the sum of the measures of the interior angles of a convex pentagon ; of a convex octagon.

#10 Find the number of sides of a convex polygon

if the sum of the measures of its angles is 900 ; 1620 ; 2340.

#11 Find the measures of each interior and exterior angle of a regular pentagon and a regular octagon.

#12 Find the measure of angle $\angle a$ for the following :

(1)

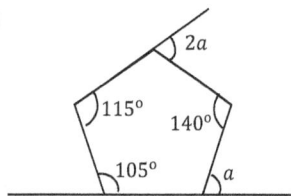

(2)

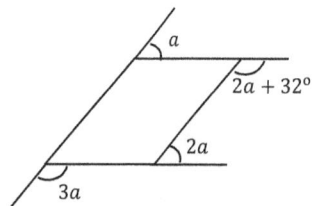

Quadrilaterals

Chapter 3   Quadrilaterals

3-1   Quadrilaterals

1.   Definition
2.   Special Quadrilaterals

3-2   Trapezoids

1.   Trapezoids

   (1) Identifying a Trapezoid

   (2) Midpoint Segment

2.   Isosceles Trapezoids

   (1) Properties of a Trapezoid

   (2) Conditions needed for a isosceles Trapezoid

3-3   Parallelograms

1.   Properties of a Parallelogram
2.   Conditions needed for a Parallelogram
3.   The areas of triangles in a Parallelogram
4.   Special Parallelograms

   (1) Rhombus

      1)   Properties of a Rhombus

      2)   Condition needed for a Rhombus

   (2) Rectangle

      1)   The Property of a Rectangle

      2)   Condition needed for a Rectangle

   (3) Square

      1)   The Property of a Square

3-4   Relationship

1.   Classification of Quadrilaterals
2.   Properties of Diagonals
3.   Areas of triangles

Chapter 3. Quadrilaterals

3-1 Quadrilaterals

1. Definition

A quadrilateral is a four-sided polygon.

For example,

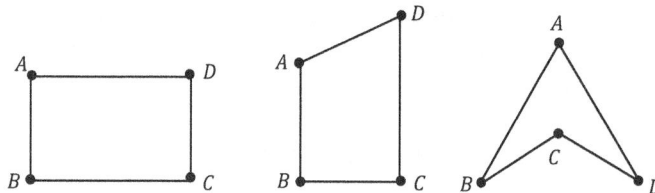

Note that :

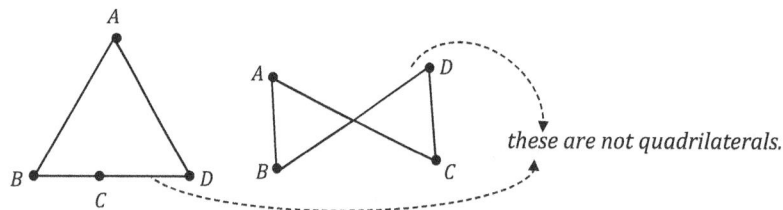

these are not quadrilaterals.

For any points *A, B, C,* and *D* :

if any three of the points do not lie on a single line and all segments $\overline{AB}, \overline{BC}, \overline{CD}$, and $\overline{DA}$ intersect at their end points, then the connection of the four segments is called a *quadrilateral.*

The four segments are called the *sides,* and the four end points *A, B, C,* and *D* are called the *vertices.*

The angles ∠ABC, ∠BCD, ∠CDA, and ∠DAB are called the *angles,* and are denoted briefly as ∠B, ∠C, ∠D, and ∠A, respectively.

The quadrilateral is denoted by ▱ABCD.

2. Special Quadrilaterals

A quadrilateral with two parallel sides is a *trapezoid*.

If the both pairs of opposite sides of a quadrilateral are parallel, then the quadrilateral is a *parallelogram*.

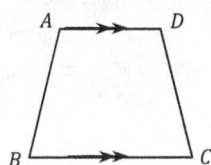
Trapezoid

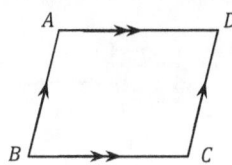
Parallelogram

A quadrilateral with four right angles is a *rectangle*.

If a rectangle has four congruent sides, then the rectangle is a *square*.

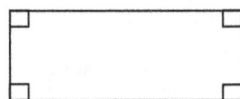
Rectangle

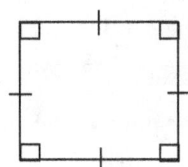
Square
(Regular polygon)

Note that :

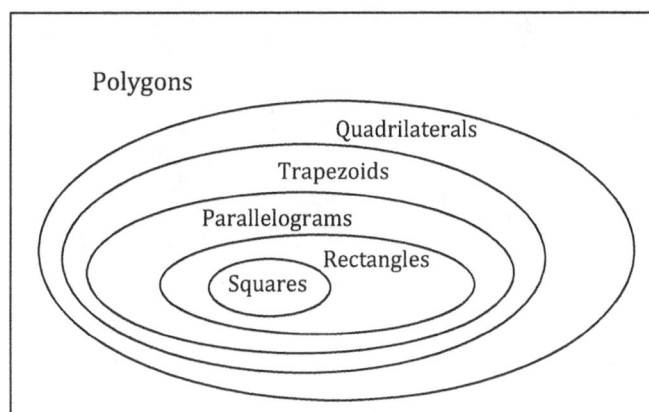

3-2 Trapezoids

1. Trapezoids

A trapezoid is a quadrilateral with only one pair of parallel sides.

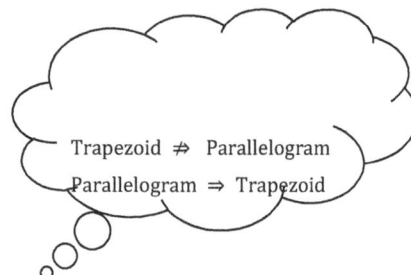

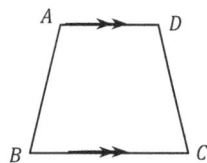

Trapezoid $\not\Rightarrow$ Parallelogram

Parallelogram $\Rightarrow$ Trapezoid

Quadrilateral
Trapezoid

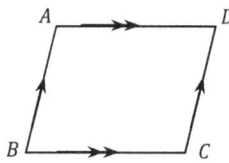

Quadrilateral
Trapezoid
Parallelogram

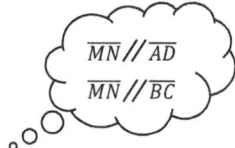

$\overline{MN} \mathbin{/\mkern-5mu/} \overline{AD}$

$\overline{MN} \mathbin{/\mkern-5mu/} \overline{BC}$

(1) Identifying a Trapezoid

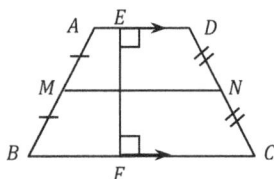

The parallel sides are the bases of a trapezoid.

Bases : $\overline{AD}$, $\overline{BC}$

The nonparallel sides are the legs of a trapezoid.

Legs : $\overline{AB}$, $\overline{DC}$

The segment connecting the midpoints of the legs is the median of a trapezoid.

Median : $\overline{MN}$

Any segment connecting the bases which is perpendicular to the bases is the altitude (height) of a trapezoid.

Altitude (height) : $\overline{EF}$

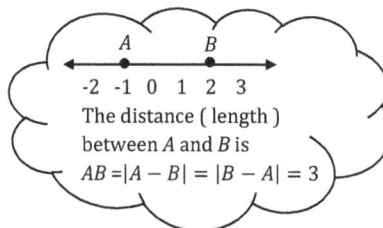

The distance (length) between A and B is $AB = |A - B| = |B - A| = 3$

(2) Midpoint Segment

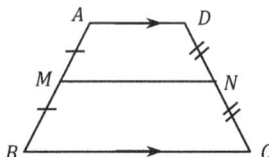

For a trapezoid $\square ABCD$,

the length of the median $\overline{MN}$ is equal to $MN = \frac{AD + BC}{2}$.

For example, if the lengths of the bases of a trapezoid are 3 and 5,

the length of the median is $MN = \frac{3+5}{2} = \frac{8}{2} = 4$.

2. Isosceles Trapezoids

Parallelogram ⇒ Isosceles Trapezoid
Isosceles Trapezoid ⇏ Parallelogram

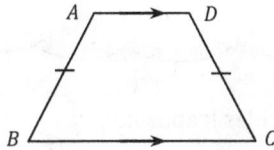

If a trapezoid $\square ABCD$ has

at least one pair of congruent opposite sides,

then the trapezoid is called an *isosceles trapezoid*.

(1) Properties of a Trapezoid

1) Base angles are congruent. $\angle A \cong \angle D$, $\angle B \cong \angle C$

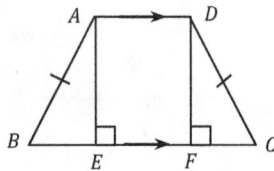

Let $\overline{AE}$ and $\overline{DF}$ be altitudes of an isosceles trapezoid $\square ABCD$.

Since $\square ABCD$ is a trapezoid, $\overline{AD} /\!/ \overline{BC}$.

Since $\square ABCD$ is isosceles, $\overline{AB} \cong \overline{DC}$.

Since $\overline{AE}$ and $\overline{DF}$ are altitudes,

$\overline{AE} \perp \overline{BC}$ and $\overline{DF} \perp \overline{BC}$; So $\overline{AE} /\!/ \overline{DF}$.

Since $\overline{AD} /\!/ \overline{EF}$ and $\overline{AE} /\!/ \overline{DF}$, $\square AEFD$ is a parallelogram.

Since the opposite sides of a parallelogram are congruent,

$\overline{AE} \cong \overline{DF}$.

By CPCTC, $\angle ABC \cong \angle DCB$; $\angle B \cong \angle C$.

Since $m \angle AEF = 90°$,

the parallelogram $\square AEFD$ is a rectangle.

So, $m \angle EAD = m \angle ADF = 90°$.

By CPCTC, $m \angle BAE \cong m \angle CDF$.

So, $m \angle BAE + m \angle EAD = m \angle CDF + m \angle FDA$.

Therefore, $m \angle BAD = m \angle CDA$; $\angle A \cong \angle D$.

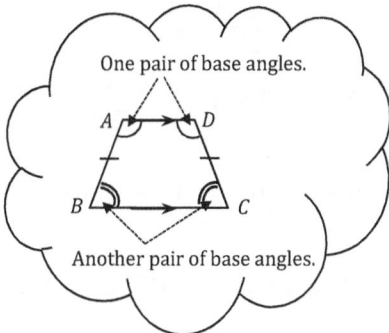

One pair of base angles.

Another pair of base angles.

CPCTC :
Corresponding parts
of congruent triangles
are congruent.

2) Opposite angles are supplementary.

$$m \angle A + m \angle C = 180^\circ, \ m \angle B + m \angle D = 180^\circ$$

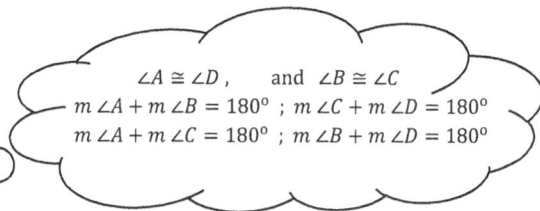

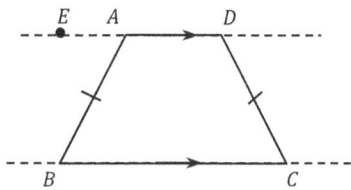

$\angle A \cong \angle D, \quad$ and $\angle B \cong \angle C$
$m \angle A + m \angle B = 180^\circ \ ; \ m \angle C + m \angle D = 180^\circ$
$m \angle A + m \angle C = 180^\circ \ ; \ m \angle B + m \angle D = 180^\circ$

Since $\square ABCD$ is an isosceles trapezoid ,

$\overline{AD} \ /\!/ \ \overline{BC}$ and $\overline{AB} \cong \overline{DC}$.

Note that

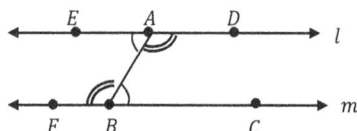

If $l \ /\!/ \ m$, then $\angle EAB \cong \angle ABC$, $\angle DAB \cong \angle ABF$ (alternate interior angles) .

Since $m \angle EAB + m \angle DAB = 180^\circ$, $m \angle DAB + m \angle ABC = 180^\circ$.

So, $m \angle A + m \angle B = 180^\circ$ and $m \angle C + m \angle D = 180^\circ$.

Since base angles of an isosceles trapezoid are congruent, $m \angle B = m \angle C$.

Therefore, $m \angle A + m \angle C = 180^\circ$, $m \angle B + m \angle D = 180^\circ$.

3) The diagonals are congruent. $\overline{AC} \cong \overline{BD}$

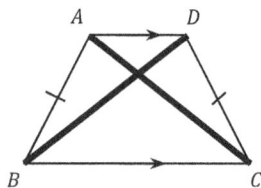

Since $\square ABCD$ is an isosceles trapezoid ,

$\overline{AD} \ /\!/ \ \overline{BC}$ and $\overline{AB} \cong \overline{DC}$.

Since the base angles of an isosceles trapezoid are congruent,

$\angle B \cong \angle C$.

Since $\overline{BC}$ is a common side of $\triangle ABC$ and $\triangle DBC$,

$\triangle ABC \cong \triangle DBC$ by SAS Postulate.

Therefore, $\overline{AC} \cong \overline{BD}$.

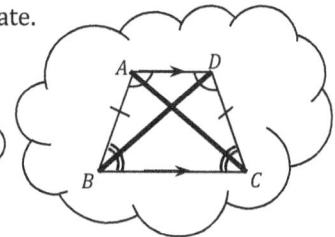

(2) Conditions needed for an isosceles Trapezoid

1) If the legs of a trapezoid are congruent, the trapezoid is isosceles.

2) If both pairs of base angles are congruent, the trapezoid is isosceles.

3) If the diagonals are congruent, the trapezoid is isosceles.

3-3 Parallelograms

$\overline{AB} /\!/ \overline{DC}$ and $\overline{AD} /\!/ \overline{BC}$
$\Rightarrow \square ABCD$ is a parallelogram.

1. Properties of a Parallelogram

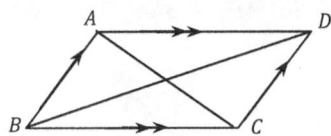

The diagonals of a parallelogram are two segments joining non-consecutive vertices.

The diagonals of $\square ABCD$ are $\overline{AC}$ and $\overline{BD}$.

(1)

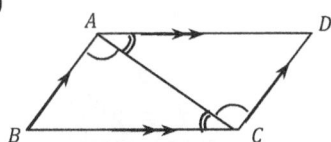

① Any two opposite sides are congruent.

$\overline{AB} \cong \overline{DC}$, $\overline{AD} \cong \overline{BC}$

② Any two opposite angles are congruent.

$\angle A \cong \angle C$, $\angle B \cong \angle D$

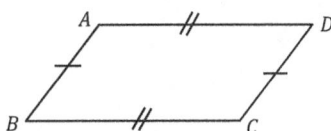

①   ②

Since $\overline{AB} /\!/ \overline{DC}$, $\angle BAC \cong \angle ACD$ (alternate interior angles)

Since $\overline{AD} /\!/ \overline{BC}$, $\angle CAD \cong \angle ACB$.

Since $\overline{AC}$ is a common segment of $\triangle ABC$ and $\triangle ACD$,

$\triangle ABC \cong \triangle ACD$ by ASA Postulate.

Therefore, $\overline{AB} \cong \overline{DC}$, $\overline{AD} \cong \overline{BC}$ and $\angle A \cong \angle C$, $\angle B \cong \angle D$.

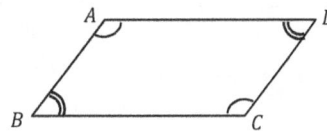

Each diagonal separates a parallelogram into two congruent triangles.

(2) Any two consecutive angles are supplementary. $m \angle A + m \angle B = 180°$; $m \angle A + m \angle D = 180°$

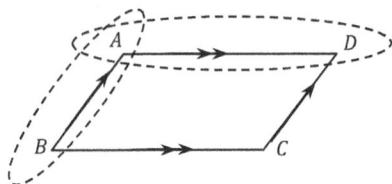

Consecutive Angles :

$\angle A$ and $\angle B$; $\angle A$ and $\angle D$

$\angle B$ and $\angle C$; $\angle C$ and $\angle D$

A midpoint divides a segment into two congruent segments.

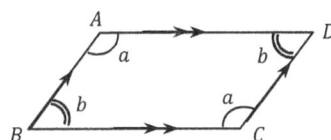

A parallelogram has two pairs of congruent opposite angles.

So, $2 m \angle a + 2 m \angle b = 360°$; $m \angle a + m \angle b = 180°$

Therefore, $m \angle A + m \angle B = m \angle A + m \angle D = 180°$

(3)   The diagonals of a parallelogram bisect each other.

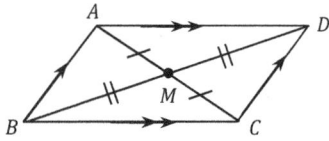

Let M be the intersecting point of two diagonals.

Since $\overline{AD} /\!/ \overline{BC}$, $\angle DAC \cong \angle BCA$ and $\angle ADB \cong \angle CBD$.

Since $\overline{AD}$ and $\overline{BC}$ are opposite sides of $\square ABCD$,

$\overline{AD}$ and $\overline{BC}$ are congruent. So $\overline{AD} \cong \overline{BC}$.

By ASA Postulate, $\triangle AMD \cong \triangle BMC$. So $\overline{AM} \cong \overline{MC}$ and $\overline{BM} \cong \overline{MD}$.

Thus, M is the midpoint of $\overline{AC}$ and $\overline{BD}$.

So, two diagonals intersect at their midpoint.

Therefore, $\overline{AC}$ bisects $\overline{BD}$ and $\overline{BD}$ bisects $\overline{AC}$.

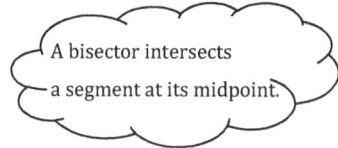

A bisector intersects a segment at its midpoint.

Note : For a given parallelogram $\square ABCD$,

if $\overline{AE}$ bisects $\angle A$, then $\triangle ABE$ is an isosceles triangle.

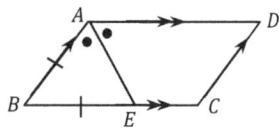

(∵ *Since $\overline{AD} /\!/ \overline{BC}$, $\angle EAD \cong \angle AEB$ by alternate interior angles.*

Therefore, $\angle BAE \cong \angle BEA$

So $\triangle ABE$ is an isosceles triangle.)

2. Conditions needed for a Parallelogram

A quadrilateral is a parallelogram if

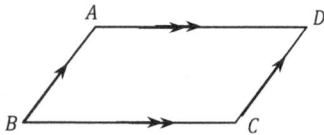

(1) both pairs of opposite sides of the quadrilateral are parallel.

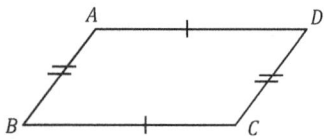

(2) both pairs of opposite sides of the quadrilateral are congruent.

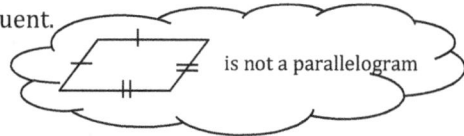

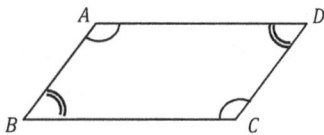

is not a parallelogram

(3) both pairs of opposite angles of the quadrilateral are congruent.

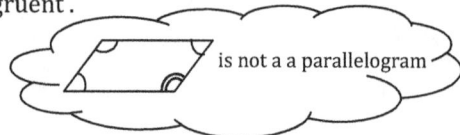

is not a a parallelogram

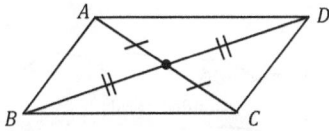

(4) the diagonals bisect each other.

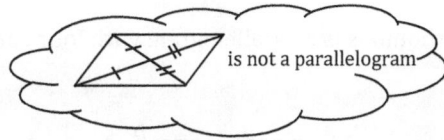

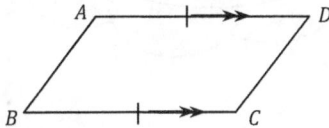

is not a parallelogram

(5) one pair of opposite sides is parallel and congruent.

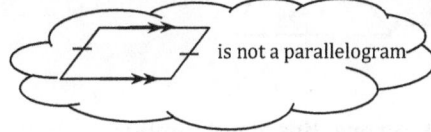

is not a parallelogram

3. The areas of triangles in a Parallelogram

The diagonals of a parallelogram bisect its area.

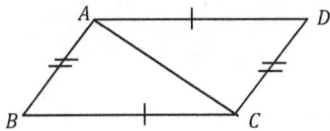

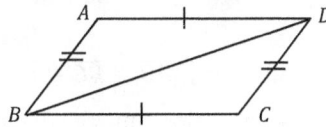

$$\triangle\, ABC \cong \triangle\, ACD \cong \triangle\, ABD \cong \triangle\, BCD$$

The diagonals divide the parallelogram into 4 triangles of equal area.

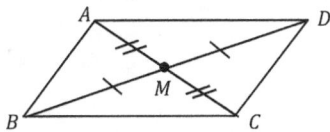

The area of $\square\, ABCD$ is equal to (area of $\triangle$ ABM) $\times\, 4$.

$$\triangle\, ABM \cong \triangle\, BCM \cong \triangle\, CDM \cong \triangle\, ADM$$

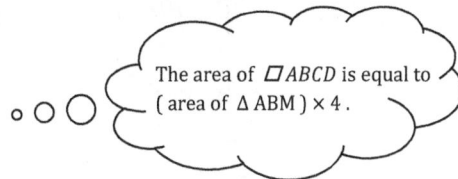

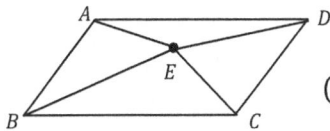

For any interior point E of $\square\, ABCD$,

$$\left(\begin{smallmatrix}\text{the area of}\\ \triangle\, ABE\end{smallmatrix}\right) + \left(\begin{smallmatrix}\text{the area of}\\ \triangle\, CDE\end{smallmatrix}\right) = \left(\begin{smallmatrix}\text{the area of}\\ \triangle\, ADE\end{smallmatrix}\right) + \left(\begin{smallmatrix}\text{the area of}\\ \triangle\, BCE\end{smallmatrix}\right)$$

$$= \frac{1}{2} \times (\text{ the area of } \square\, ABCD\,) .$$

4. Special Parallelograms

(1) Rhombus

A rhombus is a parallelogram with four congruent sides.

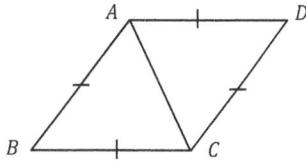

$$\overline{AB} \cong \overline{BC} \cong \overline{CD} \cong \overline{DA}$$

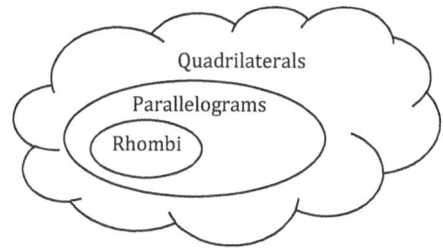

1) Properties of a Rhombus

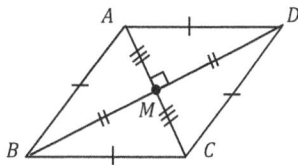

① The diagonals of a rhombus are perpendicular to each other.
$\overline{AC} \perp \overline{BD}$

(∵ Since $\square ABCD$ is a rhombus, $\overline{AB} \cong \overline{BC} \cong \overline{CD} \cong \overline{DA}$.

Since a rhombus is a parallelogram, the diagonals bisect each other.

Thus, $\overline{AM} \cong \overline{MC}$, $\overline{BM} \cong \overline{MD}$

By SSS Postulate, $\triangle ABM \cong \triangle BCM \cong \triangle CDM \cong \triangle ADM$

Since $m \angle AMB + m \angle AMD = 180°$, $m \angle AMB = m \angle AMD = 90°$.

Therefore, $\overline{AC} \perp \overline{BD}$.)

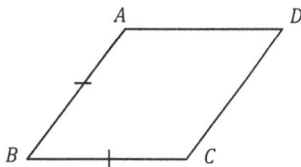

② If one pair of consecutive segments is congruent, then the parallelogram is a rhombus.

(∵ Since $\square ABCD$ is a parallelogram, $\overline{AB} \cong \overline{DC}$, $\overline{AD} \cong \overline{BC}$.

Since $\overline{AB} \cong \overline{BC}$ (given), $\overline{AB} \cong \overline{BC} \cong \overline{CD} \cong \overline{AD}$

Therefore, $\square ABCD$ is a rhombus.)

2) Condition needed for a Rhombus

If the diagonals of a quadrilateral bisect each other and are perpendicular, then the quadrilateral is a rhombus.

($\because$ If $\overline{AC} \perp \overline{BD}$, $\overline{AM} \cong \overline{MC}$, and $\overline{BM} \cong \overline{MD}$,

then $\triangle ABM \cong \triangle BCM \cong \triangle CDM \cong \triangle ADM$ by SAS Postulate.

Therefore, $\overline{AB} \cong \overline{BC} \cong \overline{CD} \cong \overline{AD}$

$\therefore \square ABCD$ is a rhombus.)

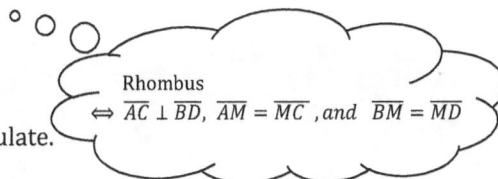

Rhombus
$\Leftrightarrow \overline{AC} \perp \overline{BD}, \overline{AM} = \overline{MC}, and \overline{BM} = \overline{MD}$

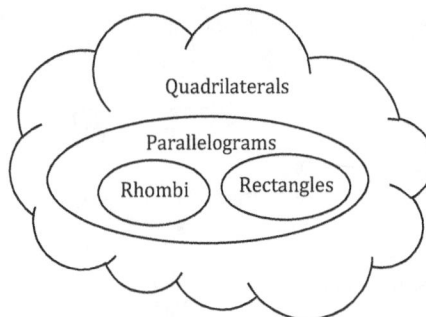

Quadrilaterals

Parallelograms

Rhombi   Rectangles

(2) Rectangle

A rectangle is a parallelogram with four right angles.

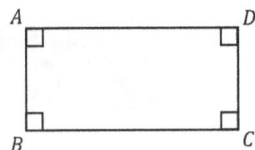

$m \angle A = m \angle B = m \angle C = m \angle D = 90^\circ$

1) The Property of a Rectangle

The diagonals of a rectangle are congruent to each other.

$\overline{AC} \cong \overline{BD}$, $\overline{AM} \cong \overline{BM} \cong \overline{CM} \cong \overline{DM}$

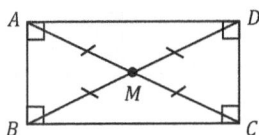

($\because$ Since $\square ABCD$ is a rectangle, $m \angle A = m \angle C = 90^\circ$.

Since $\square ABCD$ is a parallelogram, the opposite sides are congruent. $\overline{AB} \cong \overline{DC}$

Since $\overline{BC}$ is a common side of $\triangle ABC$ and $\triangle DCB$, $\triangle ABC \cong \triangle DCB$ by SAS Postulate.

Therefore, $\overline{AC} \cong \overline{BD}$.

Since a rectangle is a parallelogram, the diagonals bisect each other.

$\overline{AM} \cong \overline{MC}$, $\overline{BM} \cong \overline{MD}$

Since $\overline{AC} \cong \overline{BD}$, $\overline{AM} \cong \overline{BM} \cong \overline{CM} \cong \overline{DM}$)

2) Condition needed for a Rectangle

If one of the angles of a parallelogram is a right angle, then the parallelogram has four right angles. Therefore, the parallelogram is a rectangle.

(3) Square

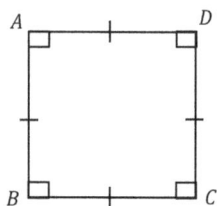

A square is a rectangle with four congruent sides.

Or, a square is a rhombus with four right angles.

1) The Property of a Square

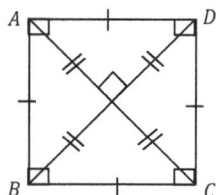

The diagonals of a square are congruent to each other.

Also, the diagonals are perpendicular and bisect each other.

Note that :

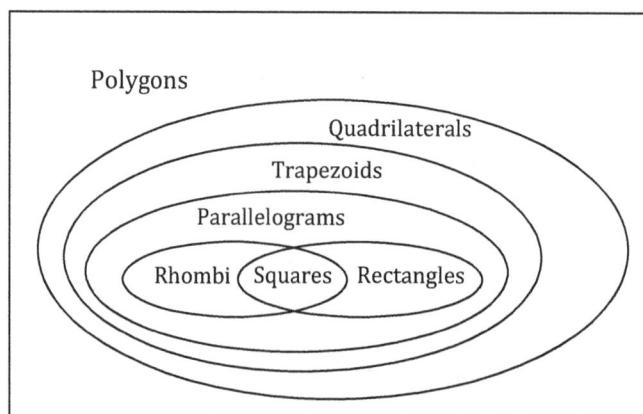

3-4 Relationship

1. Classification of Quadrilaterals

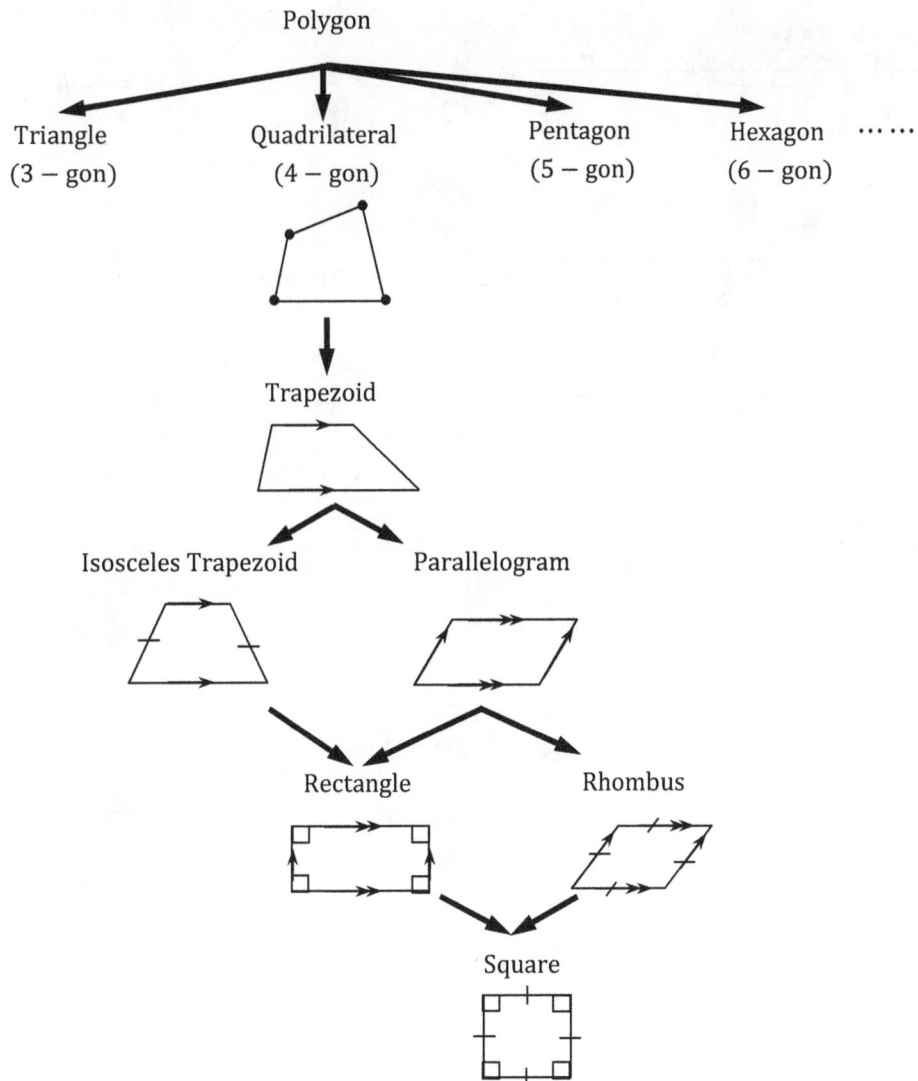

2. Properties of Diagonals

(1) Isosceles Trapezoid ⇒ Congruent diagonals.

(2) Parallelogram ⇒ Bisect each other.

(3) Rectangle ⇒ Congruent and bisect each other.

(4) Rhombus ⇒ Perpendicular and bisect each other.

(5) Square ⇒ Congruent, perpendicular, and bisect each other.

3. Areas of triangles

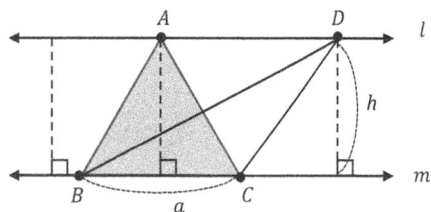

① $l \mathbin{/\mkern-5mu/} m$

⇒ The area of $\triangle ABC$ = The area of $\triangle DBC$

$= \frac{1}{2} \times BC \times \text{height} = \frac{1}{2}ah$

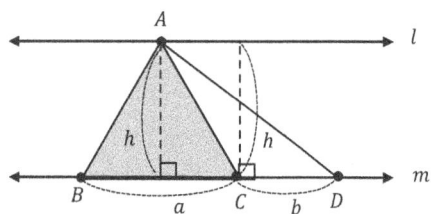

② The area of $\triangle ABC$: The area of $\triangle ACD$

$= \frac{1}{2}ah : \frac{1}{2}bh = a : b$

Exercises

#1 Determine whether the following statements are true or false.

    (1) A square is a rectangle.

    (2) A rhombus is a square.

    (3) A rectangle is a trapezoid.

    (4) A square is an isosceles trapezoid.

    (5) A square is a parallelogram.

    (6) A rhombus is an isosceles trapezoid.

    (7) A rectangle is an isosceles trapezoid.

    (8) A rectangle is a square.

    (9) A square is a rhombus.

    (10) A square is a trapezoid.

#2

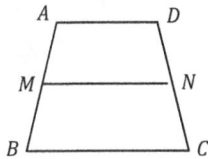

For trapezoid $\square ABCD$, the length of the median $\overline{MN}$ is 10 and the length of $\overline{BC}$ is 2 greater than twice the length of $\overline{AD}$. Find the lengths of the bases.

#3 For the following isosceles trapezoids $\square ABCD$, find the measure of $\angle a$.

(1)

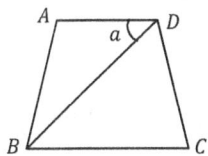

$m\angle C = 75°, \ m\angle ABD = 25°$

(2)

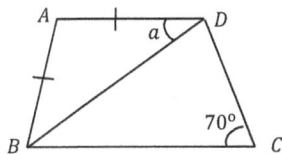

$m\angle C = 70°, \ \overline{AB} \cong \overline{AD}$

(3)

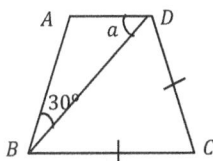

$m\angle ABD = 30°, \ \overline{BC} \cong \overline{CD}$

#4 For the following isosceles trapezoids $\square ABCD$, answer the following :

(1)

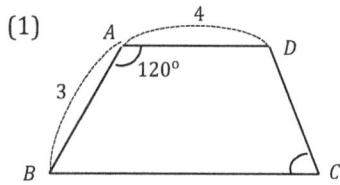

Find the length of the side $\overline{BC}$.

(2)

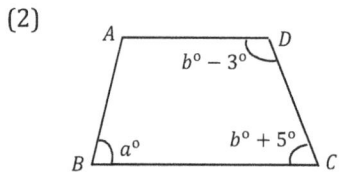

Find the values of a and b.

#5 For the following parallelograms $\square ABCD$, answer the following :

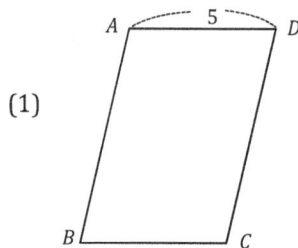

(1)

The perimeter of $\square ABCD$ is 30.

The length of $\overline{AD}$ is 5. Find the length of $\overline{AB}$.

(2)

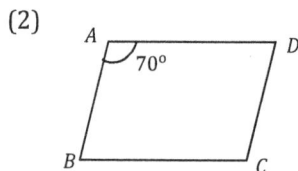

$m \angle A = 70°$.

Find the measures of $\angle C$ and $\angle D$.

(3)

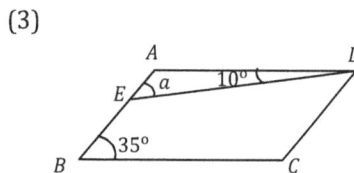

$m \angle B = 35°$, $m \angle ADE = 10°$.

Find $m \angle a$.

(4)

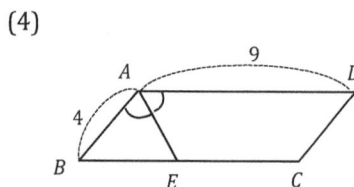

$\angle BAE \cong \angle DAE$

The lengths of $\overline{AD}$ and $\overline{AB}$ are 9 and 4, respectively.
Find the length of $\overline{EC}$.

(5)

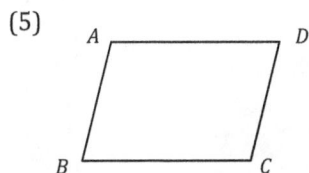

$\angle A : \angle B = 3 : 2$

Find the measure of $\angle C$.

(6)

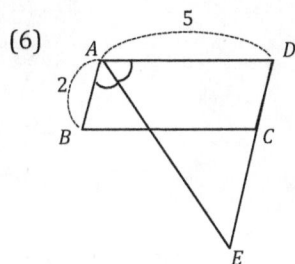

$\angle BAE \cong \angle EAD$

Find the length of $\overline{CE}$.

#6 For the parallelogram $\square ABCD$, answer the following :

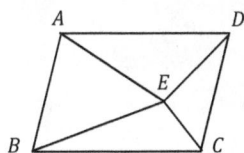

(1) The area of $\triangle ABE$ is 15 $in.^2$ The area of $\triangle BCE$ is 10 $in.^2$

The area of $\triangle CDE$ is 8 $in.^2$ Find the area of $\triangle ADE$.

(2) The area of $\square ABCD$ is 50 $in.^2$ Find the sum of the area of $\triangle ABE$ and the area of $\triangle CDE$.

(3) The area of $\square ABCD$ is 80 $in.^2$ The area of $\triangle ABE$ is 30 $in.^2$ Find the area of $\triangle CDE$.

#7 For the following rhombi $\square ABCD$, answer the following :

(1)

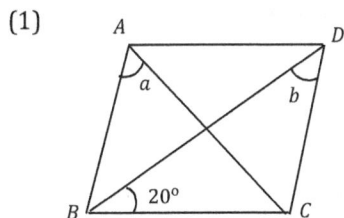

$m \angle DBC = 20°$

Find the measures of $\angle BAD$ and $\angle BDC$.

(2)

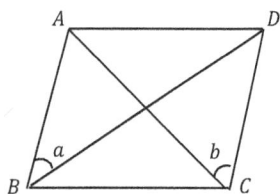

Find the sum of $m\angle a + m\angle b$.

(3)

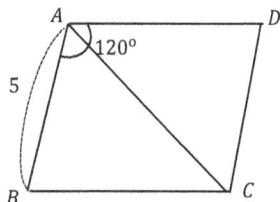

The length of $\overline{AB}$ is 5 and $m\angle A = 120°$.
Find the length of $\overline{AC}$.

#8   For the following rectangles $\square ABCD$, answer the following :

(1)

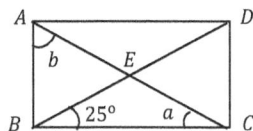

Find $m\angle a$ and $m\angle b$.

(2)

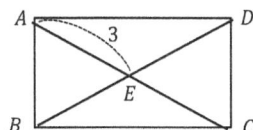

The length of $\overline{AE}$ is 3 inches.
Find the length of $\overline{BD}$.

(3)

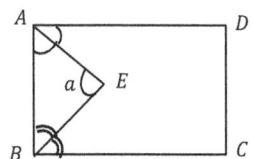

$\angle BAE \cong \angle DAE$ and $\angle ABE \cong \angle EBC$.
Find $m\angle a$.

#9 For the following squares $\square ABCD$, answer the following :

(1)

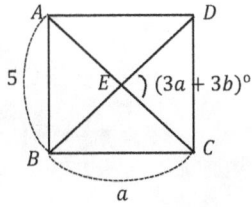

The length of $\overline{AB}$ is 5.

For the length a of $\overline{BC}$,

find the value of b when $m \angle DEC = (3a + 3b)^\circ$

(2)

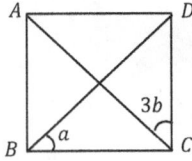

Find the measures of $\angle a$ and $\angle b$.

(3)

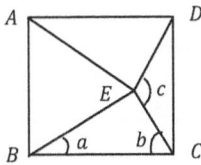

$\overline{AB} \cong \overline{AE} \cong \overline{BE}$.

Find the measures of $\angle a$, $\angle b$, and $\angle c$.

#10 Determine whether the following statements are true or false.

(1) A rhombus with congruent diagonals is a square.

(2) A parallelogram with one right angle is a rectangle.

(3) The diagonals of a rectangle are perpendicular to each other.

(4) A parallelogram is a rectangle.

(5) A parallelogram with diagonals perpendicular to each other is a rhombus.

Circles

Chapter 4 Circles

4-1 Basic Concepts of Circles

1. Definition
   (1) Circle
   (2) Radius
   (3) Chord
   (4) Secant
   (5) Diameter
   (6) Tangent
2. Tangent Lines to circles
   (1) Internally Tangent Circles
   (2) Externally Tangent Circles
   (3) Common Internal Tangents
   (4) Common External Tangents

4-2 Circumscribed and Inscribed Circles

1. Circumscribed Circle
   (1) Definition
      1) Centroid
      2) Circumscribed Circle
      3) Circumcenter
   (2) The Property of Circumcenters
   (3) Applications
   (4) The Location of the Circumcenter
2. Inscribed Circle
   (1) Definition
      1) Incenter
      2) Inscribed Circle
   (2) The Property of Incenters
   (3) Applications
   (4) The Location of the Incenter

4-3 Arcs of Circles

1. Definition
   (1) Central Angle
   (2) Arc $\widehat{AB}$
2. Measures of Arcs
   (1) Measures of Arcs
      1) Measuring Minor Arcs
      2) Measuring Major Arcs
      3) Measuring Semicircles
   (2) Arc Addition
   (3) The Relationship between an Inscribed Angle and its Intercepted Arc
      1) Inscribed Angles
      2) Intercepted Arcs
      3) Measuring Intercepted Arcs

4-4 Arcs, Chords, Secants, and Tangents

1. Arcs and Chords
   (1) Congruent Arcs
   (2) Arcs and Chords
   (3) Bisecting Chords and Arcs
   (4) Congruent Chords
2. Chords, Secants, and Tangents

4-5 Circumscribed and Inscribed Quadrilaterals

1. Circumscribed Quadrilaterals
   (1) Properties of Circumscribed Quadrilaterals about a Circle
2. Inscribed Quadrilaterals
   (1) Properties of Inscribed Quadrilaterals in a Circle
   (2) Inscribed Quadrilaterals
   (3) Conditions for an Inscribed Quadrilateral in a Circle

CHAPTER 4

Chapter 4. Circles

A circle is the boundary of a round region in a plane (2-dimentional figure) and a sphere is the surface of a ball in space (3-dimentional figure).

4-1 Basic Concepts of Circles

1. Definition

(1) Circle

A *circle* is the set of all points of a plane at the same distance from a fixed point, called the *center*.

(2) Radius

A *radius* is a segment connecting the center and a point on a circle. All radii of a circle are congruent.

(3) Chord

A *chord* is a segment connecting two points on a circle.

A chord is a line segment but a secant is a line containing a chord.

(4) Secant

A *secant* is a line which intersects a circle in two points. A secant contains a chord.

(5) Diameter

A *diameter* is a chord through the center of a circle. A diameter is twice as long as a radius.

(6) Tangent

A *tangent* to a circle is a line which intersects the circle at exactly one point.

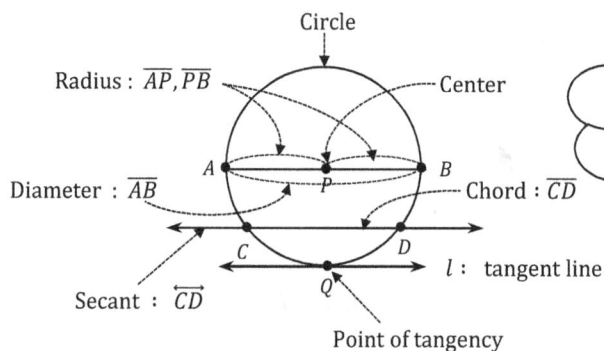

Radius : $\overline{AP}, \overline{PB}$

Circle

Center

Diameter : $\overline{AB}$

Chord : $\overline{CD}$

Secant : $\overleftrightarrow{CD}$

l : tangent line

Point of tangency

A circle has a tangent at each of its points.

2. Tangent Lines to circles

If two circles are tangent to the same line at the same point, they are tangent circles.

(1) Internally Tangent Circles

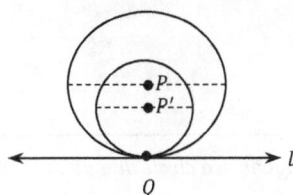

Two circles are *internally tangent* if their centers are on the same side of the tangent.

(2) Externally Tangent Circles

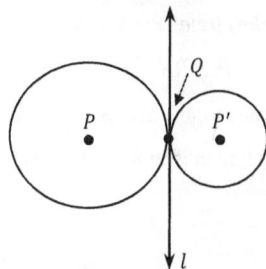

Two circles are *externally tangent* if their centers are on opposite sides of the tangent.

(3) Common Internal Tangents

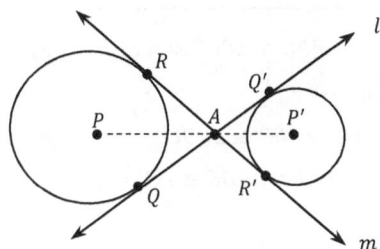

The common internal tangents l and m are tangents to each of the circles. l and m intersect the segment $\overline{PP'}$ of the centers at a point A between the centers.

(4) Common External Tangents

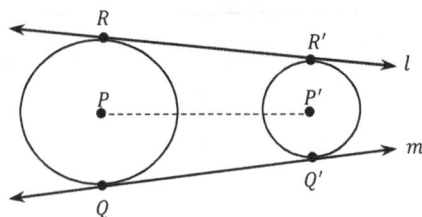

The common external tangents l and m are tangents to the same line at the same point. l and m do not intersect the segment $\overline{PP'}$ of the centers at a point between the centers.

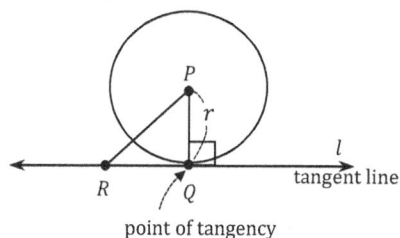

> **Note 1.** *If a line is perpendicular to a radius drawn to the point of tangency, then the line is tangent to the circle.* $\overleftrightarrow{PQ} \perp \overleftrightarrow{RQ}$

Every tangent to a circle is perpendicular to the radius drawn to the point of tangency.

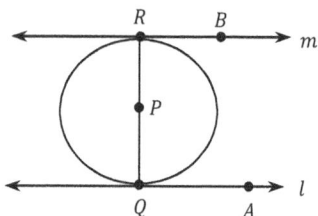

> **Note 2.** *If a line l is tangent to a circle at a point Q and l // m, then m is tangent to the circle at a point R.*

(∵ *Since* $\overline{PQ}$ *is a radius of the circle,* $\overline{PQ} \perp \overleftrightarrow{QA}$.
Since $l // m$, $\overline{RQ} \perp \overline{RB}$. *So* $\overline{RP} \perp \overline{RB}$.
Since $\overline{RP}$ *is also a radius of the circle and* $\overleftrightarrow{RB}$ *is perpendicular to the radius* $\overline{RP}$ *at point R, the line m is tangent to the circle at point R.*)

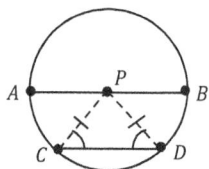

> **Note 3.** *Let* $\overline{AB}$ *be a diameter of a circle with center P. If* $\overline{AB} // \overline{CD}$, *then* $\angle APC \cong \angle PCD \cong \angle PDC \cong \angle BPD$.

(∵ *Since* $\overline{AB} // \overline{CD}$, $\angle APC \cong \angle PCD$ *and* $\angle BPD \cong \angle PDC$, *by alternate interior angles.*
Since $\overline{PC} \cong \overline{PD}$ *(radii),* $\angle PCD \cong \angle PDC$
Therefore, $\angle APC \cong \angle PCD \cong \angle PDC \cong \angle BPD$.)

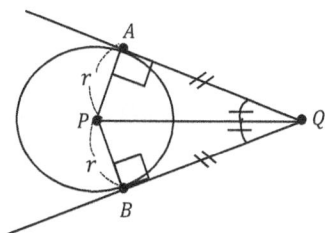

> **Note 4.** *Let* $\overline{QA}$ *and* $\overline{QB}$ *be tangents to a circle at points A and B, respectively. Then* $\overline{QA} \cong \overline{QB}$ *and* $\angle AQP \cong \angle BQP$.

(∵ *Since* $\overline{PA}$ *and* $\overline{PB}$ *are radii of a circle,* $\overline{PA} \cong \overline{PB}$.
Since every tangent to a circle is perpendicular to the radius drawn to a point of tangency, $\overline{PA} \perp \overline{QA}$ *and* $\overline{PB} \perp \overline{QB}$
i.e. $m\angle PAQ = m\angle PBQ = 90°$
Since $\overline{PQ}$ *is the common hypotenuse of* $\triangle APQ$ *and* $\triangle BPQ$,
$\triangle APQ \cong \triangle BPQ$ *by HL Theorem.*
Therefore, $\overline{QA} \cong \overline{QB}$ *and* $\angle AQP \cong \angle BQP$.)

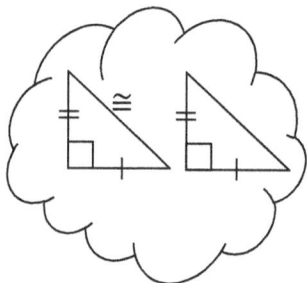

4-2 Circumscribed and Inscribed Circles

1. Circumscribed Circle

(1) Definition

1) Centroid (Center of gravity)

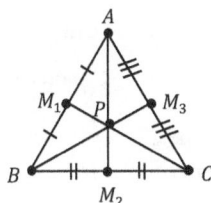

Let M_1, M_2, and M_3 be the three midpoints of $\overline{AB}$, $\overline{BC}$, and $\overline{AC}$, respectively. The medians $\overline{AM_2}$, $\overline{BM_3}$, and $\overline{CM_1}$ are formed by connecting three vertices and the midpoints of the opposite sides. The intersection point P of the three medians is called the *centroid* or *center of gravity* of the triangle.

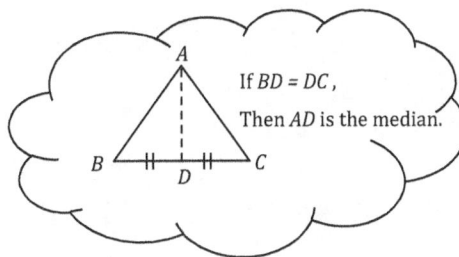

If $BD = DC$,
Then AD is the median.

The length of the segment connecting the centroid and the vertex is $\frac{2}{3}$ the length of the median.

$$AP : PM_2 = BP : PM_3 = CP : PM_1 = 2 : 1$$

2) Circumscribed Circle

A *circumscribed circle* is a circle passing through the three vertices of a triangle.

The center of the circumscribed circle is also the centroid of the triangle.

3) Circumcenter

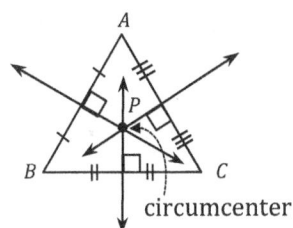

circumcenter

Three perpendicular bisects of the sides of a triangle must intersect in a single point.

This point of concurrency (intersection) is called the *circumcenter* of the triangle.

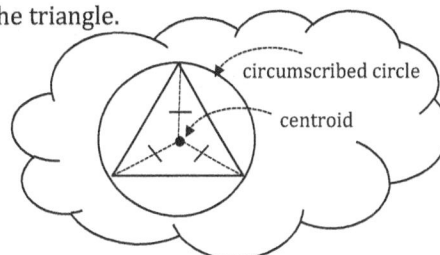

circumscribed circle

centroid

A circle is a set of all points on a plane at the same distance from a fixed point, called the center.

Radius segment

Center

Tangent line

Point of tangency

(2) The Property of Circumcenters

$$AP = BP = CP$$

The lengths from the circumcenter to the vertices of the triangle are the same.

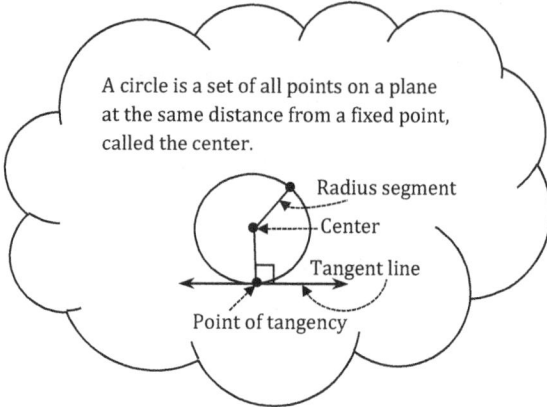

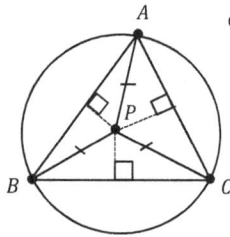

Since the centroid P is the center of the circumscribed circle, $\overline{AP}$, $\overline{BP}$, and $\overline{CP}$ are radii of the circle.

Therefore, $AP = BP = CP$

Three perpendicular bisects of the sides of a triangle intersect in a single point.

(3) Applications

1) $m \angle a + m \angle b + m \angle c = 90º$

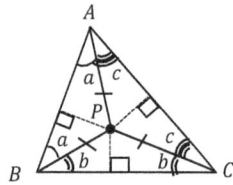

Since $AP = BP = CP$,

all triangles $\triangle PAB, \triangle PBC,$ and $\triangle PAC$

are isosceles.

$$2 m \angle a + 2m \angle b + 2m \angle c = 180º$$

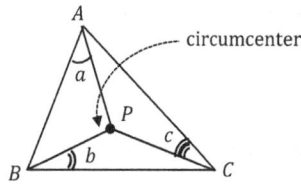

circumcenter

Therefore,

$$m \angle a + m \angle b + m \angle c = 90º$$

2) $m \angle BPC = 2m \angle A$

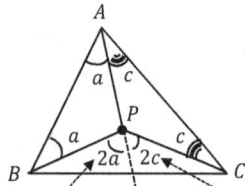

circumcenter

Since $m \angle A = m \angle a + m \angle c$,

$$2m \angle BPC = 2m \angle A$$

∵ an exterior angle of $\triangle ABP$ ∵ an exterior angle of $\triangle APC$

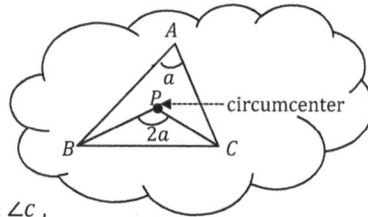

3) **Δ ADP ≅ Δ BDP, Δ BEP ≅ Δ CEP, Δ CFP ≅ Δ AFP** by SAS Postulate.

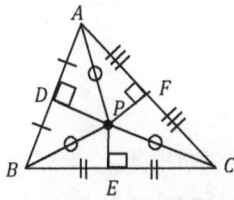

(4) The Location of the Circumcenter

The point P is the circumcenter of ΔABC .

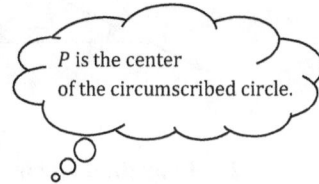

P is the center of the circumscribed circle.

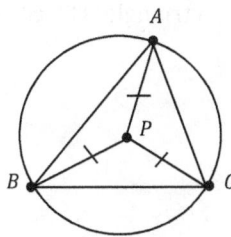
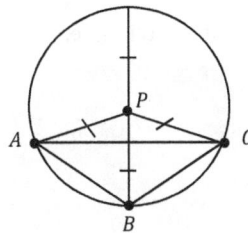
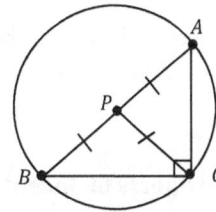

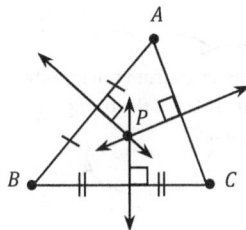
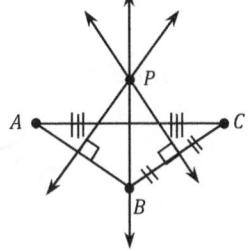
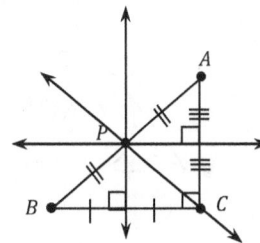

| | | |
| Acute | Obtuse | Right |

1) If the angles of a triangle are acute, then the circumcenter is located in the interior of the triangle.

2) If a triangle has an obtuse angle, then the circumcenter is located in the exterior of the triangle.

3) If a triangle is right, then the circumcenter is the midpoint of the hypotenuse.

2. Inscribed Circle

(1) Definition

1) Incenter

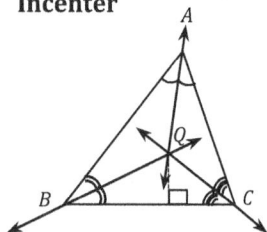

Three angle bisectors of a triangle must intersect in a single point. The intersection point is called the *incenter* of the triangle.

2) Inscribed Circle

An inscribed circle is a circle which is tangent to all sides of a triangle. The center of the inscribed circle is the incenter of the triangle.

(2) The Property of Incenters

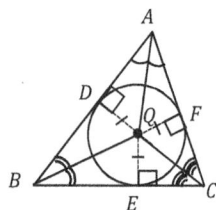

$$QD = QE = QF$$

The lengths from the incenter to the sides of the triangle are the same.

The circle is inscribed in $\triangle ABC$.

Since the incenter Q is the center of the inscribed circle, $\overline{DQ}$, $\overline{EQ}$, and $\overline{FQ}$ are radii of the circle.

Therefore, $DQ = EQ = FQ$.

(3) Applications

1) Perimeter of an isosceles triangle

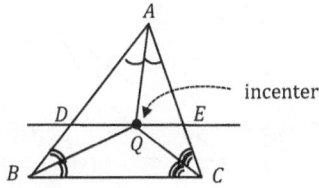

The point Q is the incenter of $\triangle ABC$.

If $\overline{DE} \, /\!/ \, \overline{BC}$, then $\triangle DBQ$ and $\triangle ECQ$ are isosceles triangles and the perimeter of $\triangle ADE$ is $m\,\overline{AB} + m\,\overline{AC}$.

($\because$ If $\overline{DE} \, /\!/ \, \overline{BC}$, then $\angle DQB \cong \angle QBC$ and $\angle EQC \cong \angle QCB$ by alternate internal angles.

So $\triangle DBQ$ and $\triangle ECQ$ are isosceles.

Since $m\,\overline{DE} = m\,\overline{DQ} + m\,\overline{QE} = m\,\overline{DB} + m\,\overline{EC}$,

$$m\,\overline{AD} + m\,\overline{AE} + m\,\overline{DE} = (m\,\overline{AD} + m\,\overline{AE}) + (m\,\overline{DB} + m\,\overline{EC})$$
$$= (m\,\overline{AD} + m\,\overline{DB}) + (m\,\overline{AE} + m\,\overline{EC})$$
$$= m\,\overline{AB} + m\,\overline{AC} \;)$$

2) $m\angle a + m\angle b + m\angle c = 90^\circ$

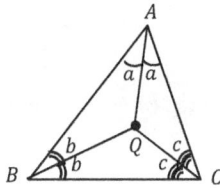

Since Q is the incenter of the triangle $\triangle ABC$,

$2\,m\angle a + 2m\angle b + 2m\angle c = 180^\circ$.

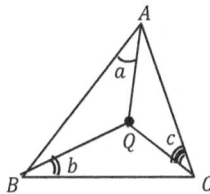

Therefore, $m\angle a + m\angle b + m\angle c = 90^\circ$.

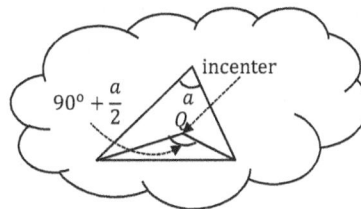

3) $m\angle BQC = 90^\circ + \dfrac{1}{2}\, m\angle A$

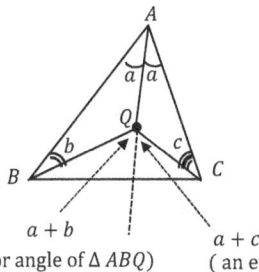

$m\angle BQC$

$= (m\angle a + m\angle b) + (m\angle a + m\angle c)$

$= (m\angle a + m\angle b + m\angle c) + m\angle a$

$= 90^\circ + \dfrac{1}{2}\, m\angle A$

$a + b$
(an exterior angle of $\triangle ABQ$)

$a + c$
(an exterior angle of $\triangle AQC$)

Note : For any triangle △ABC, the altitude (height) of the triangle is the length of the segment which is perpendicular to the opposite side, called the base, from one vertex.

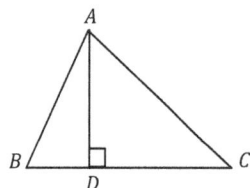

Altitude : AD
Base : $\overline{BC}$

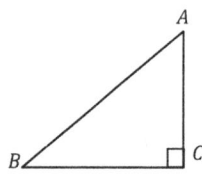

Altitude : AC
Base : $\overline{BC}$

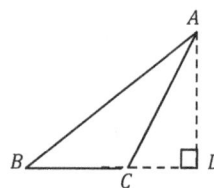

Altitude : AD
Base : $\overline{BC}$

For example, the altitude of an isosceles triangle is the length of the median of the triangle.

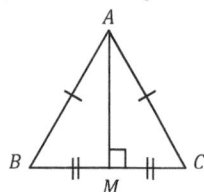

AM is the altitude.
$\overline{BC}$ *is the base.*

In other *words, altitude is the perpendicular distance from the base to the opposite vertex.*

4)

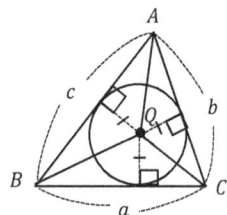

Let the lengths of three sides of △ABC be a, b, and c, respectively.

If the inscribed circle has a radius length r, then the area of the triangle is $\frac{1}{2}r(a+b+c)$.

∵ Since △ABC = △QAB + △QBC + △QCA ,

the area of the triangle △ABC is $\frac{1}{2}rc + \frac{1}{2}ra + \frac{1}{2}rb = \frac{1}{2}r(a+b+c)$

5)

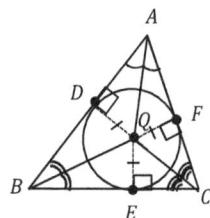

The inscribed circle is tangent to all sides of the triangle △ABC .

Let the points of tangency be D, E, and F, respectively.

Then $\overline{AD} \cong \overline{AF}$, $\overline{BD} \cong \overline{BE}$, and $\overline{CE} \cong \overline{CF}$.

∵ By RHA correspondence, △ADQ ≅ △AFQ , △BDQ ≅ △BEQ, and △CEQ ≅ △CFQ .

Therefore, $\overline{AD} \cong \overline{AF}$, $\overline{BD} \cong \overline{BE}$, and $\overline{CE} \cong \overline{CF}$.

(4) The Location of the Incenter

The point Q is the incenter of $\triangle ABC$.

1)

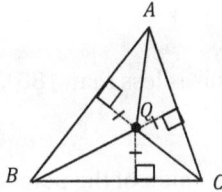

For any triangle,

the incenter is located in the interior of the triangle.

2)

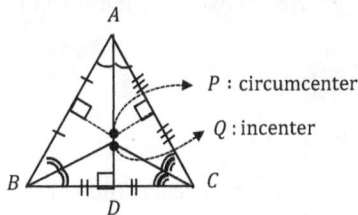

P : circumcenter

Q : incenter

If a triangle $\triangle ABC$ is isosceles,

then the incenter and circumcenter are located on

the bisector $\overline{AD}$.

3)

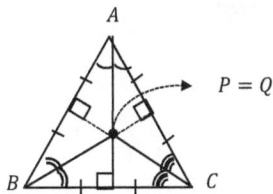

$P = Q$

If a triangle $\triangle ABC$ is regular,

then the circumcenter and incenter are the same point.

4-3 Arcs of Circles

1. Definition

(1) Central Angle

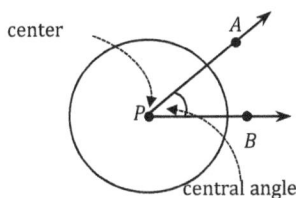

center

central angle

A *central angle* is an angle with a vertex at the center
of a circle.

$\angle APB$ is a central angle of a circle with center P.

(2) Arc $\widehat{AB}$

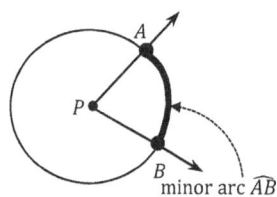

For a circle with center P, the *minor arc* $\widehat{AB}$ is the union of points A, B which are on the circle but not the end points of a diameter and all points of the circle that locate in the interior of $\angle APB$.

The minor arc $\widehat{AB}$ measures less than $180°$.

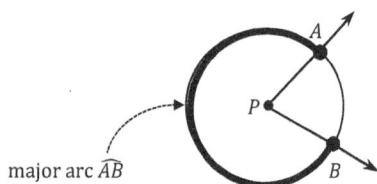

The *major arc* $\widehat{AB}$ is the union of the points A, B and all points of the circle that locate in the exterior of $\angle APB$.

The major arc $\widehat{AB}$ measures greater than $180°$.

The two points A and B are called the *end points* of the arc $\widehat{AB}$.

If the two points A and B are the end points of a diameter $\overline{AB}$ of a circle, then arcs $\widehat{ACB}$ and $\widehat{ADB}$ are semicircles . Each semicircle $\widehat{AB}$ is the union of the points A, B, and the points of the circle that locate in the given half-plane with segment $\overline{AB}$ as its edge.

The semicircles $\widehat{ACB}$ and $\widehat{ADB}$ measures $180°$.

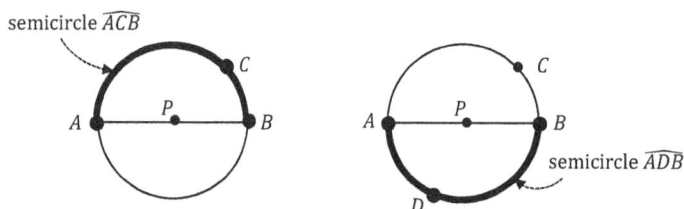

The end points of semicircles are also the end points of a diameter.

2. Measures of Arcs

(1) Measures of Arcs

The measure of an arc $\widehat{AB}$ is denoted by $m\,\widehat{AB}$.

Note : The measure of an arc is defined as the measure of the central angle with sides that intersect the endpoints of the arc. The measure of an arc does not depend on the size of the circle.
All corresponding arcs have the same measure.

Note :

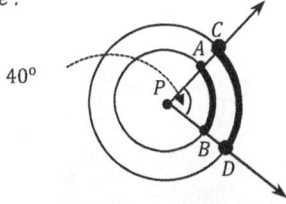

$$m\,\widehat{AB} = m\,\widehat{CD} = 40°$$

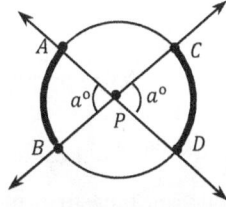

$$m\,\widehat{AB} = m\,\widehat{CD} = a°$$
$$\because \angle APB \cong \angle CPD \ (\text{vertical angles})$$

For a circle, the longer the arc becomes, the greater the measure of its angle.

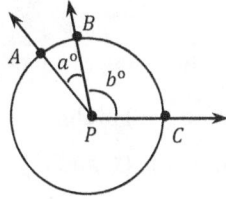

If $m\,\widehat{AB} < m\,\widehat{BC}$, then $a < b$.

1) Measuring Minor Arcs

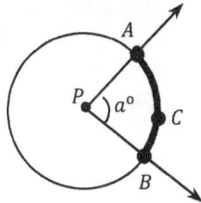

$$m\,\widehat{ACB} = a°$$

The measure of a minor arc $\widehat{ACB}$

is the measure of the corresponding central angle.

2) Measuring Major Arcs

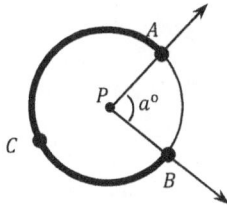

$$m\,\widehat{ACB} = 360° - a°$$

The measure of a major arc $\widehat{ACB}$ is equal to 360° minus

the measure of the corresponding minor arc.

The measure of a major arc is greater than 180° and less than 360°.

3) Measuring Semicircles

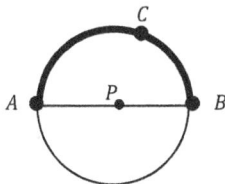

$$m\,\widehat{ACB} = 180°$$

The measure of a semicircle is 180°.

(2) Arc Addition

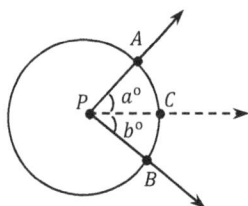

If the point C is on $\widehat{AB}$, then $m\,\widehat{ACB} = m\widehat{AC} + m\widehat{CB}$.

The measure of a whole is equal to the sum of the measures of its parts.

(3) The Relationship between an Inscribed Angle and its Intercepted Arc

1) Inscribed Angles

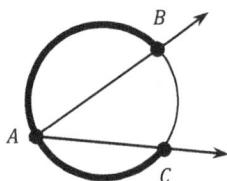

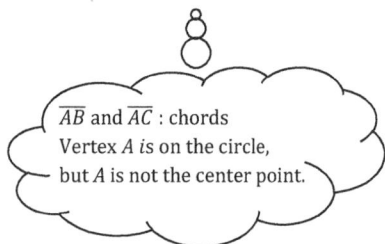

An angle, $\angle BAC$, is inscribed in an arc $\widehat{BAC}$ if

① the sides ($\overrightarrow{AB}$ and $\overrightarrow{AC}$) of the angle $\angle BAC$ contain the end points (B and C) of the arc $\widehat{BC}$, and

② the vertex A of the angle is a point of the arc, but the point is not an end point.

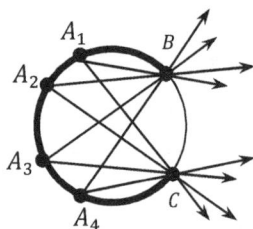

$\overline{AB}$ and $\overline{AC}$: chords
Vertex A is on the circle, but A is not the center point.

For example,
$\angle BA_1C$, $\angle BA_2C$, $\angle BA_3C$, $\angle BA_4C$, $\cdots$
are all inscribed angles in the arc $\widehat{BC}$.

2) Intercepted Arcs

An angle intercepts an arc if

① the sides of the angle contain the end points of the arc, and

② all the other points of the arc are in the interior of the angle.

$\angle BA_iC$ intercepts $\widehat{BC}$, $i = 1, 2, 3, 4$

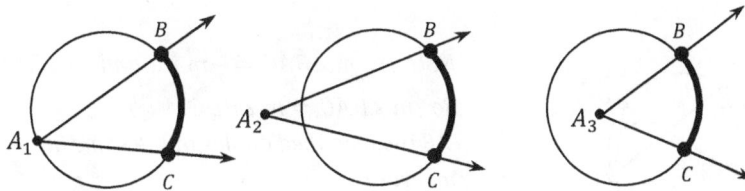

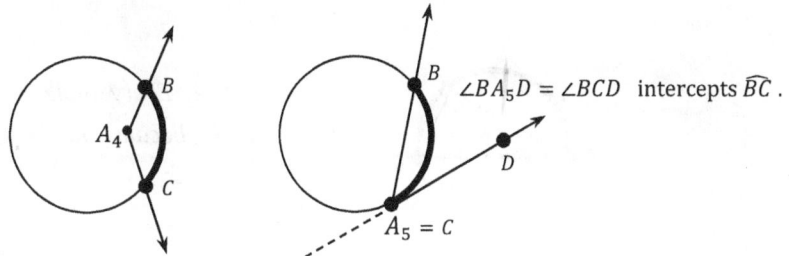

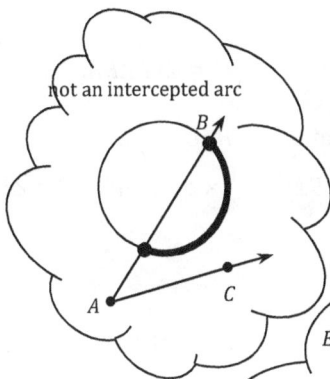

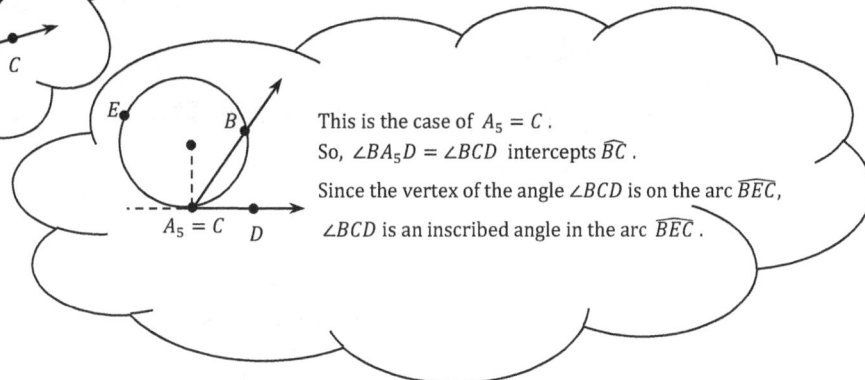

$\angle BA_5D = \angle BCD$ intercepts $\widehat{BC}$.

not an intercepted arc

This is the case of $A_5 = C$.

So, $\angle BA_5D = \angle BCD$ intercepts $\widehat{BC}$.

Since the vertex of the angle $\angle BCD$ is on the arc $\widehat{BEC}$,

$\angle BCD$ is an inscribed angle in the arc $\widehat{BEC}$.

3) Measuring Intercepted Arcs

The measure of an inscribed angle of a circle is half the measure of its intercepted arc.

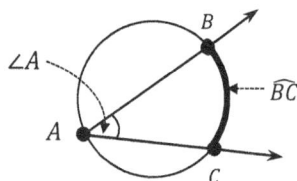

If $\angle A$ is inscribed in an arc $\widehat{BAC}$ of a circle, intercepting the arc $\widehat{BC}$,

then $m \angle A = \frac{1}{2} m \widehat{BC}$.

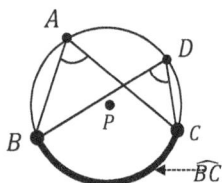

Note 1: $m \angle BAC = \frac{1}{2} m \widehat{BC}$ and $m \angle BDC = \frac{1}{2} m \widehat{BC}$

So $m \angle BAC = m \angle BDC$.

All the inscribed angles in the same arc have the same measures.

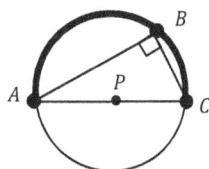

Note 2: Every angle inscribed in a semicircle is a right angle, because $m \angle ABC = \frac{1}{2} \times 180° = 90°$.

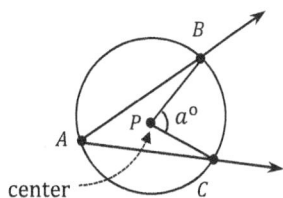

Note 3: $m \angle BAC = \frac{1}{2} \times a°$; $a° = 2m \angle BAC$

$(\because a° = m \widehat{BC}$ and $m \angle BAC = \frac{1}{2} \widehat{BC}$

Therefore, $m \angle BAC = \frac{1}{2} \times a°$; $a° = 2m \angle BAC$)

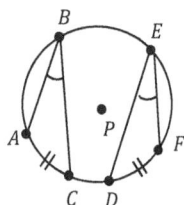

Note 4: If $m \widehat{AC} = m \widehat{DF}$, then $m \angle ABC = m \angle DEF$

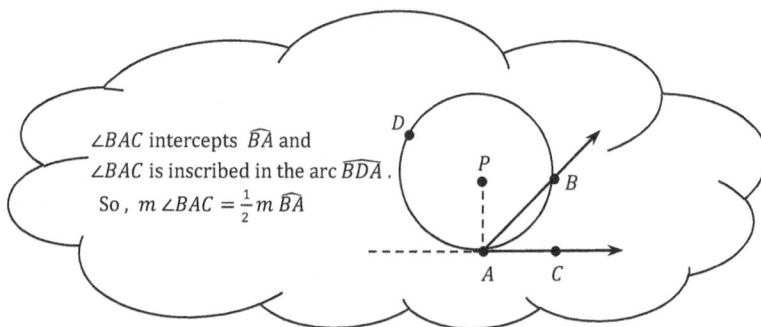

$\angle BAC$ intercepts $\widehat{BA}$ and $\angle BAC$ is inscribed in the arc $\widehat{BDA}$.

So, $m \angle BAC = \frac{1}{2} m \widehat{BA}$

4-4 Arcs, Chords, Secants, and Tangents

> In a circle, the degree measure of an arc is equal to the measure of the central angle that intercepts the arc.

1. Arcs and Chords

(1) Congruent Arcs

The arcs in the same circle or congruent circles are called *congruent* if they have the same measure.

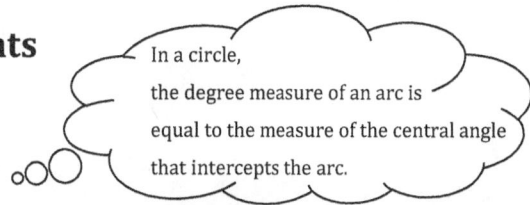

(2) Arcs and Chords

Congruent chords intercept congruent arcs.

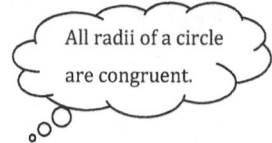

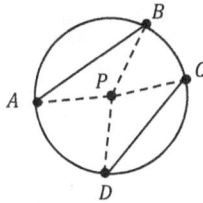

> All radii of a circle are congruent.

Let $\overline{AB} \cong \overline{CD}$.

Since $\overline{AP} \cong \overline{BP} \cong \overline{CP} \cong \overline{DP}$ (radii),

$\triangle ABP \cong \triangle CDP$ by SSS Postulate.

So, $m\angle APB = m\angle CPD$.

Since the measure of the central angle is the measure of the corresponding arc,

$m\widehat{AB} = m\widehat{CD}$ So, $\widehat{AB} \cong \widehat{CD}$

Therefore, if $\overline{AB} \cong \overline{CD}$, then $\widehat{AB} \cong \widehat{CD}$.

Conversely, if two arcs are congruent, then their corresponding chords are congruent.

Note : In the same circle or congruent circles, congruent central angles have congruent chords.

Conversely, congruent chords have congruent central angles.

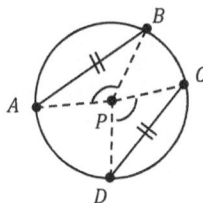

$\angle APB \cong \angle CPD \Leftrightarrow \overline{AB} \cong \overline{CD}$

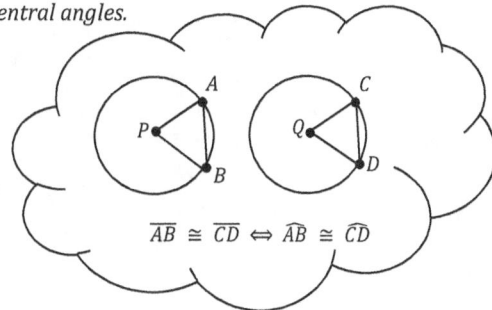

$\overline{AB} \cong \overline{CD} \Leftrightarrow \widehat{AB} \cong \widehat{CD}$

(3) Bisecting Chords and Arcs

1) A perpendicular segment from the center of a circle to a chord bisects the chord.

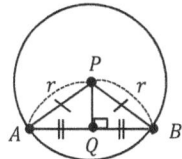

($\because$ Since $\overline{PA}$ and $\overline{PB}$ are radii of a circle, $\overline{PA} \cong \overline{PB}$.

Then $\triangle PAB$ is isosceles. So $\angle PAB \cong \angle PBA$.

Since $m\angle PQA = m\angle PQB = 90°$, $\triangle PAQ \cong \triangle PBQ$ by SAA Postulate or HL Theorem.

So, $\overline{AQ} \cong \overline{BQ}$ Thus, Q is the midpoint of $\overline{AB}$. Therefore, $\overline{PQ}$ bisects $\overline{AB}$.)

2) If $\overset{\frown}{AB} \cong \overset{\frown}{AC}$ and $\overline{AQ}$ bisects $\angle BAC$, then $\overline{AQ}$ bisects $\overset{\frown}{BQC}$.

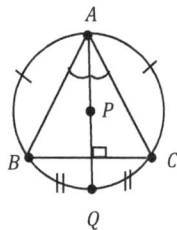

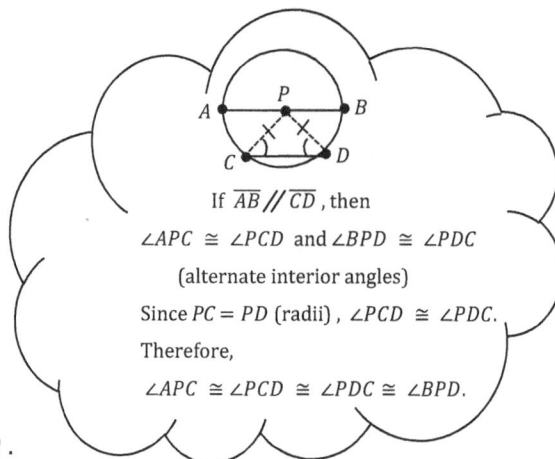

If $\overline{AB} /\!/ \overline{CD}$, then

$\angle APC \cong \angle PCD$ and $\angle BPD \cong \angle PDC$

(alternate interior angles)

Since $PC = PD$ (radii), $\angle PCD \cong \angle PDC$.

Therefore,

$\angle APC \cong \angle PCD \cong \angle PDC \cong \angle BPD$.

($\because$ Since $\overset{\frown}{AB} \cong \overset{\frown}{AC}$, $\overline{AB} \cong \overline{AC}$.

So $\triangle ABC$ is isosceles.

Since $\overline{AQ}$ bisects $\angle BAC$, $m \angle BAQ = m \angle CAQ$.

Since $m \angle BAQ = \frac{1}{2} m \overset{\frown}{BQ}$ and $\angle CAQ = \frac{1}{2} m \overset{\frown}{CQ}$, $\frac{1}{2} m \overset{\frown}{BQ} = \frac{1}{2} m \overset{\frown}{CQ}$; $m \overset{\frown}{BQ} = m \overset{\frown}{CQ}$

Thus, Q is the midpoint of $\overset{\frown}{BQC}$.

Therefore, $\overline{AQ}$ bisects $\overset{\frown}{BQC}$.)

(4) Congruent Chords

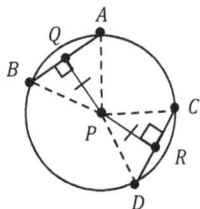

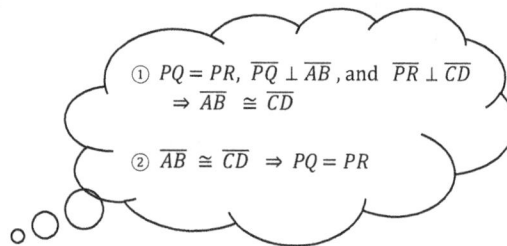

① $PQ = PR$, $\overline{PQ} \perp \overline{AB}$, and $\overline{PR} \perp \overline{CD}$
$\Rightarrow \overline{AB} \cong \overline{CD}$

② $\overline{AB} \cong \overline{CD}$ $\Rightarrow PQ = PR$

Let $PQ = PR$ (the same distance from a center P), $\overline{PQ} \perp \overline{AB}$, and $\overline{PR} \perp \overline{CD}$.

Since $AB = AQ + QB$, $CD = CR + RD$, and $AP = BP = CP = DP$ (radii),

by HL Theorem, $\triangle AQP \cong \triangle BQP \cong \triangle CPR \cong \triangle DPR$

Thus, $\overline{AQ} \cong \overline{CR}$ and $\overline{BQ} \cong \overline{DR}$

So, $AQ + QB = CR + RD$.

Thus, $AB = CD$ (same distance)

Therefore, chords $\overline{AB}$ and $\overline{CD}$ are congruent.

2. Chords, Secants, and Tangents

(1) $m \angle BAC = \frac{1}{2} a^\circ = \frac{1}{2} m \, \widehat{AB}$

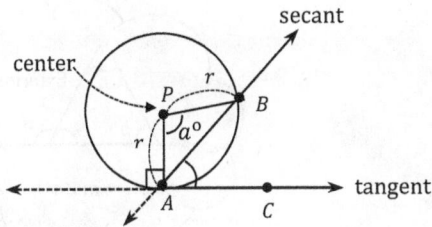

The measure of an arc is the measure of the corresponding central angle.

Therefore, $m \, \widehat{AB} = a^\circ$

So $m \angle BAC = \frac{1}{2} m \, \widehat{AB}$

If a secant and a tangent intersect at a point on a circle, then the measure of the angle formed by the secant ray and the tangent ray is half the measure of the intercepted arc.

($\because$ Since $\overrightarrow{AC}$ is a tangent ray, $m \angle PAB + m \angle BAC = 90^\circ$.

Since $\overline{PA}$ and $\overline{PB}$ are radii, $m \angle PAB = m \angle PBA$

So, $a^\circ + 2 \, m \angle PAB = 180^\circ$

Therefore, $m \angle BAC = 90^\circ - m \angle PAB = 90^\circ - \frac{180^\circ - a^\circ}{2} = \frac{1}{2} a^\circ$)

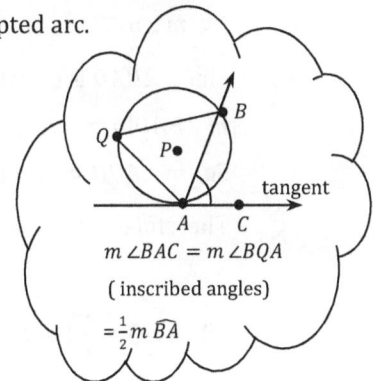

$m \angle BAC = m \angle BQA$

(inscribed angles)

$= \frac{1}{2} m \, \widehat{BA}$

(2) $m \angle AQB = \frac{1}{2} \left(m \, \widehat{AB} + m \, \widehat{CD} \right) = m \angle CQD$

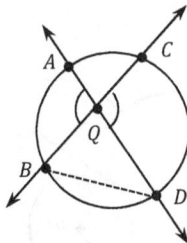

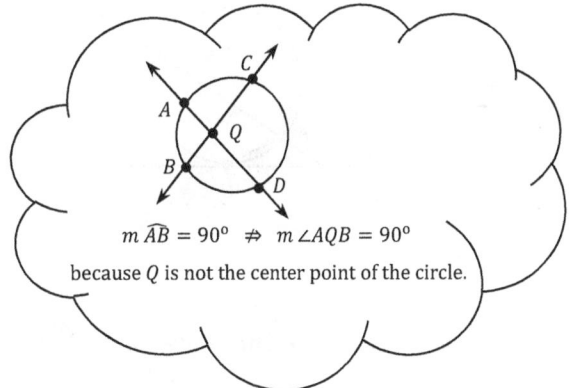

$m \, \widehat{AB} = 90^\circ \; \not\Rightarrow \; m \angle AQB = 90^\circ$

because Q is not the center point of the circle.

The measure of an angle formed by two secants intersecting inside of a circle is half the sum of the measures of the intercepted arcs.

($\because$ By the measures of the exterior angles, $m \angle AQB = m \angle QBD + m \angle QDB$.

Since $m \angle QBD = m \angle CBD = \frac{1}{2} m \, \widehat{CD}$ and $m \angle QDB = m \angle ADB = \frac{1}{2} m \, \widehat{AB}$,

$m \angle AQB = \frac{1}{2} m \, \widehat{CD} + \frac{1}{2} m \, \widehat{AB} = \frac{1}{2} \left(m \, \widehat{CD} + m \, \widehat{AB} \right)$.

Since $\angle AQB$ and $\angle CQD$ are vertical angles, $m \angle AQB = m \angle CQD = \frac{1}{2} \left(m \, \widehat{AB} + m \, \widehat{CD} \right)$.

(3) The measure of an angle formed by ① two secants ; ② a secant and a tangent ; ③ two tangents of a circle intersecting at an exterior point Q is half the difference of the measures of the intercepted arcs.

① $m \angle Q = \frac{1}{2} \left(m \, \widehat{BD} - m \, \widehat{AC} \right)$

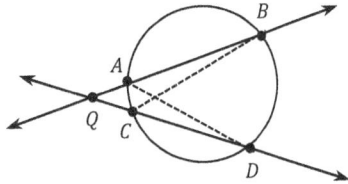

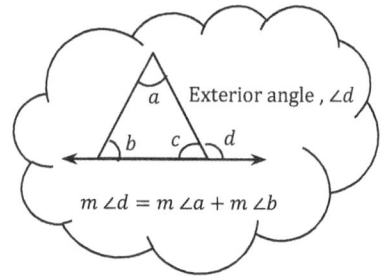

Exterior angle , $\angle d$

$m \angle d = m \angle a + m \angle b$

($\because m \angle BAD = \frac{1}{2} m \, \widehat{BD}$, $m \angle ADC = \frac{1}{2} m \, \widehat{AC}$.

Since $\angle BAD$ is an exterior angle of $\Delta \, AQD$,

$m \angle BAD = m \angle AQD + m \angle ADQ$

So, $m \angle AQD = m \angle BAD - m \angle ADQ$; $m \angle Q = m \angle BAD - m \angle ADC$

Therefore, $m \angle Q = \frac{1}{2} m \, \widehat{BD} - \frac{1}{2} m \, \widehat{AC} = \frac{1}{2} \left(m \, \widehat{BD} - m \, \widehat{AC} \right)$)

② $m \angle Q = \frac{1}{2} \left(m \, \widehat{BC} - m \, \widehat{AC} \right)$

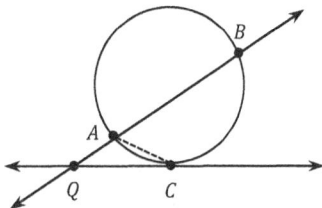

($\because m \angle BAC = \frac{1}{2} m \, \widehat{BC}$, $m \angle ACQ = \frac{1}{2} m \, \widehat{AC}$.

Since $m \angle BAC = m \angle Q + m \angle ACQ$,

$m \angle Q = m \angle BAC - m \angle ACQ$

$= \frac{1}{2} m \, \widehat{BC} - \frac{1}{2} m \, \widehat{AC}$

$= \frac{1}{2} \left(m \, \widehat{BC} - m \, \widehat{AC} \right)$)

③ $m \angle Q = \frac{1}{2} \left(m \, \widehat{ADB} - m \, \widehat{ACB} \right)$

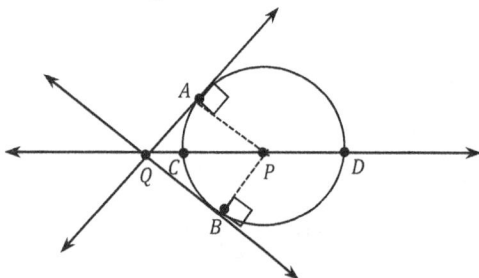

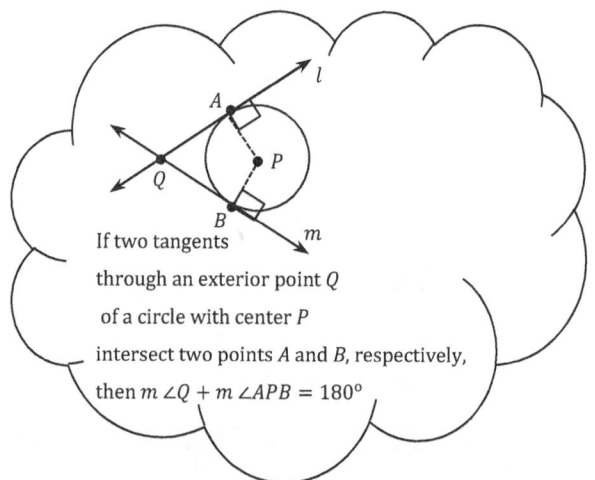

If two tangents through an exterior point Q of a circle with center P intersect two points A and B, respectively, then $m \angle Q + m \angle APB = 180°$

$(\because m \angle Q = m \angle AQP + m \angle BQP.$

Since $m \angle AQD = \frac{1}{2}\left(m \, \widehat{AD} - m \, \widehat{AC}\right)$ and $m \angle BQP = \frac{1}{2}\left(m \, \widehat{DB} - m \, \widehat{CB}\right),$

$m \angle Q = \frac{1}{2}\left(m \, \widehat{AD} - m \, \widehat{AC}\right) + \frac{1}{2}\left(m \, \widehat{DB} - m \, \widehat{CB}\right)$

$\qquad = \frac{1}{2}\left(m \, \widehat{AD} + m \, \widehat{DB}\right) - \frac{1}{2}\left(m \, \widehat{AC} + m \, \widehat{CB}\right)$

$\qquad = \frac{1}{2}m \, \widehat{ADB} - \frac{1}{2}m\widehat{ACB} = \frac{1}{2}\left(m \, \widehat{ADB} - m \, \widehat{ACB}\right)$)

(4) $QA \cdot QD = QB \cdot QC$

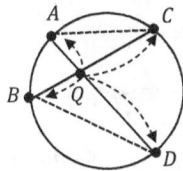

AA Similarity Theorem

$\angle A \cong \angle D$ and $\angle B \cong \angle E$,
then $\triangle ABC \cong \triangle DEF$

If two chords $\overline{AD}$ and $\overline{BC}$ of the same circle intersect at a point Q, then $QA \cdot QD = QB \cdot QC$

$\triangle ABQ \sim \triangle CDQ$

$QA : QC = QB : QD$

$\dfrac{QA}{QC} = \dfrac{QB}{QD}$

$\therefore QA \cdot QD = QB \cdot QC$

① $3 : 2 = 6 : x$; $\dfrac{3}{2} = \dfrac{6}{x}$; $3x = 12$; $x = 4$

② $3 : 6 = 2 : x$; $\dfrac{3}{6} = \dfrac{2}{x}$; $3x = 12$; $x = 4$

③ $3 \cdot x = 2 \cdot 6$; $3x = 12$; $x = 4$

Note :

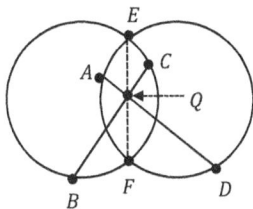

$QB \cdot QC = QE \cdot QF = QA \cdot QD$

$QA \cdot QD = QB \cdot QC$

(5) $QA \cdot QB = QC \cdot QD$

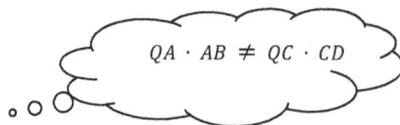

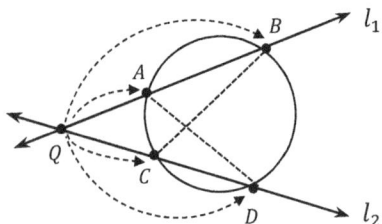

$QA \cdot AB \neq QC \cdot CD$

If a secant l_1 through an exterior point Q intersects a circle at points A and B, and another secant l_2 through the same exterior point Q intersects the circle at points C and D, then $QA \cdot QB = QC \cdot QD$

(∵ Since $\angle QBC$ and $\angle QDA$ are inscribed in the same arcs $\overarc{ABC}$ and $\overarc{ADC}$, $\angle QBC \cong \angle QDA$. Since $\angle Q$ is a common angle in $\triangle QBC$ and $\triangle QDA$, $\triangle QBC \sim \triangle QDA$ by AA similarity Theorem.

So, $QB : QD = QC : QA$

$\dfrac{QB}{QD} = \dfrac{QC}{QA}$

Therefore, $QA \cdot QB = QC \cdot QD$)

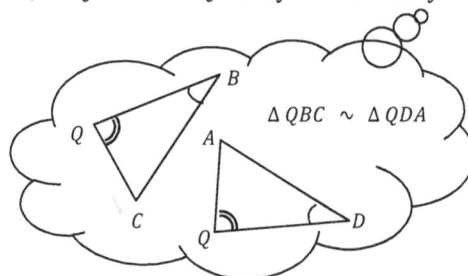

$\triangle QBC \sim \triangle QDA$

Note :

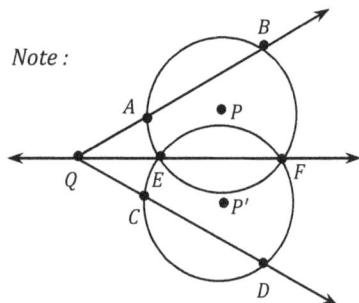

$QA \cdot QB = QE \cdot QF = QC \cdot QD$

$QA \cdot QB = QC \cdot QD$

(6) $QA \cdot QB = QC^2$; $QC = \sqrt{QA \cdot QB}$

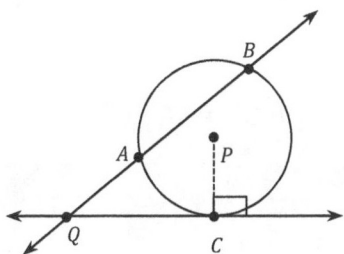

If a secant through an exterior point Q intersects a circle at points A and B, and a tangent through the same exterior point Q intersects the circle at a point C, then $QA \cdot QB = QC^2$

$QC^2 = (QC)^2$

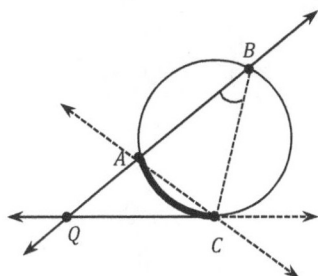

(∵ Since $\overset{\frown}{AC}$ is the arc intercepted by $\angle ABC$ and $\angle ACQ$,

$m \angle QBC = \frac{1}{2} m \overset{\frown}{AC}$ and $m \angle QCA = \frac{1}{2} m \overset{\frown}{AC}$.

So, $\angle QBC \cong \angle QCA$.

Since $\angle Q$ is a common angle in ΔQAC and ΔQBC , $\Delta QAC \sim \Delta QBC$ by AA similarity Theorem.

So, $QA : QC = QC : QB$; $\frac{QA}{QC} = \frac{QC}{QB}$

Therefore, $QA \cdot QB = QC \cdot QC$ ∴ $QA \cdot QB = QC^2$)

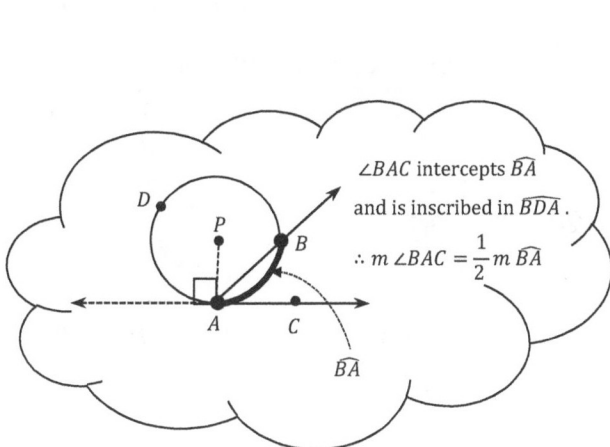

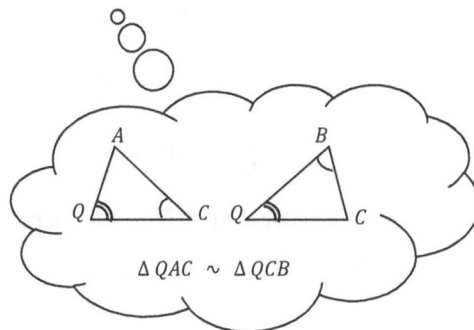

$\angle BAC$ intercepts $\overset{\frown}{BA}$ and is inscribed in $\overset{\frown}{BDA}$.
∴ $m \angle BAC = \frac{1}{2} m \overset{\frown}{BA}$

$\Delta QAC \sim \Delta QCB$

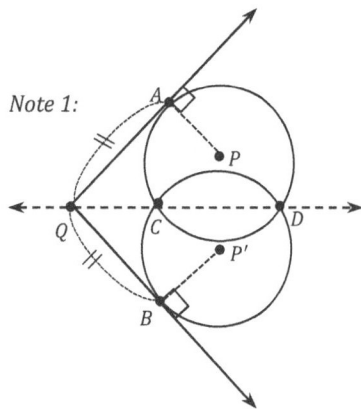

Note 1:

Since $QA^2 = QC \cdot QD$ and $QB^2 = QC \cdot QD$,

$QA^2 = QB^2$

$\therefore \quad QA = QB$

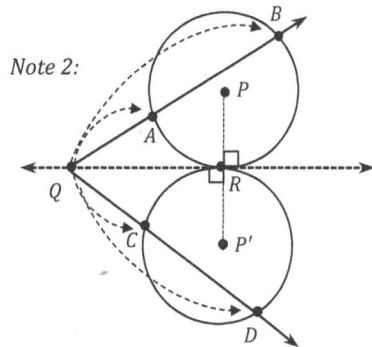

Note 2:

$QA \cdot QB = QR^2 = QC \cdot QD$

$\therefore \quad QA \cdot QB = QC \cdot QD$

(7) A diameter and a chord

① For $\overline{AD} \perp \overline{BC}$, $\quad QA \cdot QD = QB^2 = QC^2$

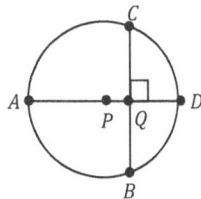

$\because QA \cdot QD = QB \cdot QC = QB^2 = QC^2$

② For a circle with radius r, $\quad QA \cdot QD = r^2 - PQ^2$

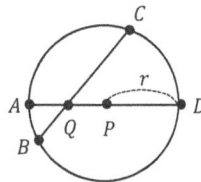

$\because QA \cdot QD = (r - PQ)(r + PQ) = r^2 - PQ^2$

③ For a circle with radius r, $\quad QA \cdot QC = QP^2 - r^2$

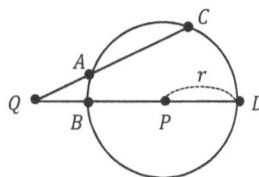

$\because QA \cdot QC = QB \cdot QD$

$\qquad = (QP - r)(QP + r)$

$\qquad = QP^2 - r^2$

(8) Two circles with a common chord,

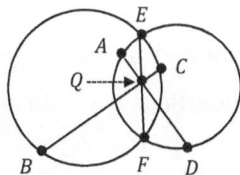

① If chords $\overline{AD}$ and $\overline{BC}$ intersect the common cord $\overline{EF}$ at a point Q, then $QA \cdot QD = QB \cdot QC$

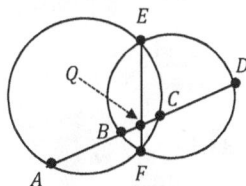

② The points A, B, C, and D all lie on a single line. If chords $\overline{AC}$ and $\overline{BD}$ intersect the common cord $\overline{EF}$ at a point Q, then $QA \cdot QC = QB \cdot QD$
(∵ Since $QA \cdot QC = QE \cdot QF$ and $QB \cdot QD = QE \cdot QF$,
  $QA \cdot QC = QB \cdot QD$)

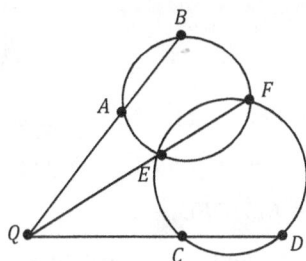

③ If chords $\overline{AB}$ and $\overline{CD}$ through an exterior point Q intersect the common chord $\overline{EF}$ through the same exterior point Q, then $QA \cdot QB = QC \cdot QD$
(∵ Since $QA \cdot QB = QE \cdot QF$ and $QE \cdot QF = QC \cdot QD$,
  $QA \cdot QB = QC \cdot QD$)

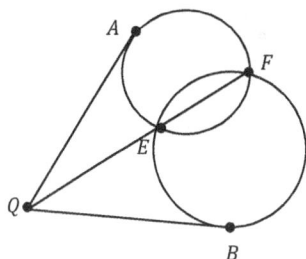

④ If two tangents through an exterior point Q intersect the common chord $\overline{EF}$ through the same exterior point Q, then $QA = QB$.
(∵ Since $QA^2 = QE \cdot QF$ and $QB^2 = QE \cdot QF$,
  $QA^2 = QB^2$.
Since QA and QB are lengths and length must be greater than or equal to zero, $QA = QB$.)

(9) Two circles with a common tangent.

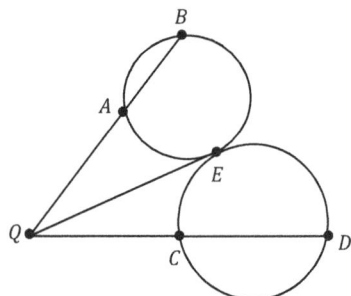

If chords $\overline{AB}$ and $\overline{CD}$ through an exterior point Q intersect the common tangent line, then $QA \cdot QB = QC \cdot QD$

($\because$ Since $QE^2 = QA \cdot QB$ and $QE^2 = QC \cdot QD$,

$QA \cdot QB = QC \cdot QD$)

4-5 Circumscribed and Inscribed Quadrilaterals

1. Circumscribed Quadrilateral

(1) Properties of Circumscribed Quadrilaterals about a Circle

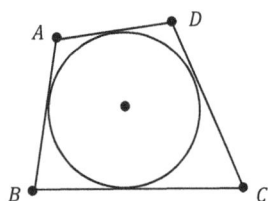

1) The sum of the lengths of one pair of opposite sides is equal to the sum of the lengths of the other pair of opposite sides. ; $AB + CD = BC + AD$

2) If $AB + CD = BC + AD$, then the quadrilateral is circumscribed about a circle.

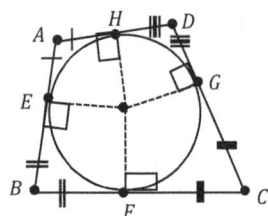

($\because$ Since $\overleftrightarrow{AB}$, $\overleftrightarrow{BC}$, $\overleftrightarrow{CD}$, and $\overleftrightarrow{AD}$ are tangent lines of the circumscribed circle,

$\overline{AE} \cong \overline{AH}$, $\overline{BE} \cong \overline{BF}$, $\overline{CF} \cong \overline{CG}$, $\overline{DG} \cong \overline{DH}$.

Therefore,

$AB + CD = AE + EB + CG + GD = AH + BF + CF + HD$

$\qquad = (AH + HD) + (BF + CF) = AD + BC$.

The converse is also true.)

Note : Circumscribed triangles about a circle

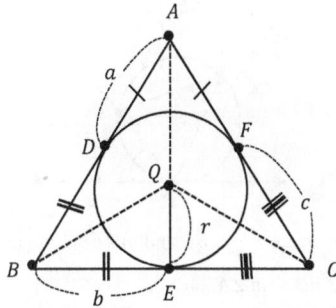

① $\overline{AD} \cong \overline{AF}$, $\overline{BD} \cong \overline{BE}$, $\overline{CE} \cong \overline{CF}$

② *The perimeter of $\triangle ABC$ is*

$$AB + BC + AC = (AD + DB) + (BE + EC) + (AF + FC) = 2\,(AD + BE + CF) = 2\,(a + b + c).$$

③ *The area of $\triangle ABC$ is the sum of the areas of $\triangle QAB$, $\triangle QBC$, and $\triangle QCA$. That is,*

$$\text{the area of } \triangle ABC \ = \tfrac{1}{2}\, r\, AB + \tfrac{1}{2}\, r\, BC + \tfrac{1}{2}\, r\, AC = \tfrac{1}{2}\, r\, (AB + BC + AC)$$

2. Inscribed Quadrilaterals

(1) Properties of Inscribed Quadrilaterals in a Circle

1) Any two opposite angles are supplementary.

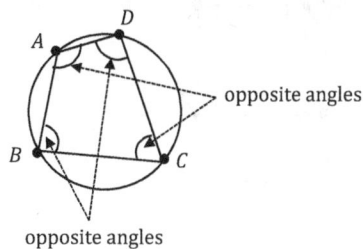

$$m \angle A + m \angle C = 180° \quad ; \quad m \angle B + m \angle D = 180°$$

2) The measure of an exterior angle at a vertex of an inscribed quadrilateral in a circle is equal to the measure of the interior opposite angle.

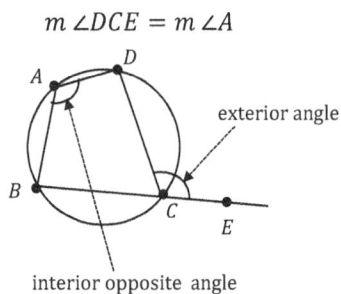

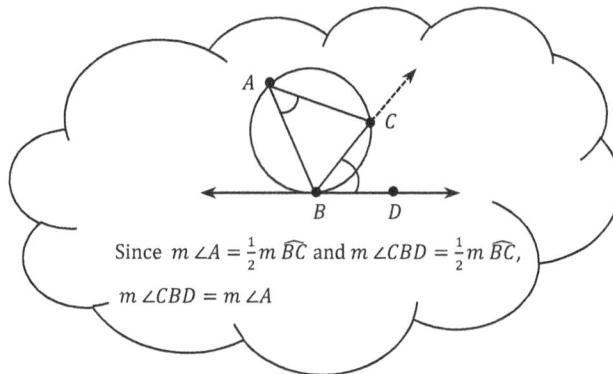

$m \angle DCE = m \angle A$

exterior angle

interior opposite angle

Since $m \angle A = \frac{1}{2} m \, \widehat{BC}$ and $m \angle CBD = \frac{1}{2} m \, \widehat{BC}$,

$m \angle CBD = m \angle A$

If the side BC is extended to E, then the exterior angle $\angle DCE$ of the inscribed quadrilateral $\square ABCD$ is equal to the interior opposite angle $\angle A$.

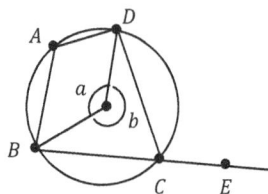

($\because$ Since $m \angle A = \frac{1}{2} m \angle b$ and $m \angle C = \frac{1}{2} m \angle a$,

$m \angle A + m \angle C = \frac{1}{2} (m \angle a + m \angle b)$.

Since $m \angle a + m \angle b = 360°$, $m \angle A + m \angle C = 180°$

Similarly, $m \angle B + m \angle D = 180°$

Since $m \angle C + m \angle DCE = 180°$,
$m \angle DCE = 180° - m \angle C = m \angle A$.)

(2) Inscribed Quadrilaterals

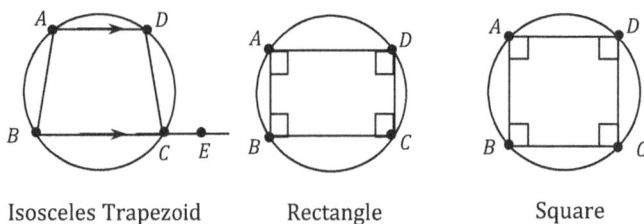

Isosceles Trapezoid   Rectangle   Square

For an isosceles trapezoid $\square ABCD$, the opposite angles are supplementary.

So, $m \angle A + m \angle C = m \angle B + m \angle D = 180°$

Since $m \angle DCB + m \angle DCE = 180°$, $m \angle DCE = m \angle A$.

Therefore, isosceles trapezoids can always be inscribed in a circle; so can rectangles and squares.

(3) Conditions for an Inscribed Quadrilateral in a Circle

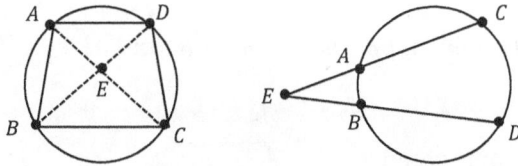

① $EA \cdot EC = EB \cdot ED$

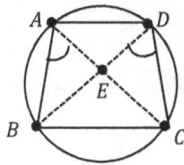

② $\angle BAC \cong \angle BDC$ ③ $m \angle B + m \angle D = 180°$ ④ $m \angle DCE = m \angle A$

1) Inscribed triangles in two circles.

Let triangles $\triangle ABC$ and $\triangle CDE$ be inscribed in two different circles.

If the circles intersect at a point C on a common tangent $\overleftrightarrow{PQ}$, then $\overline{AB} \,/\!/\, \overline{DE}$.

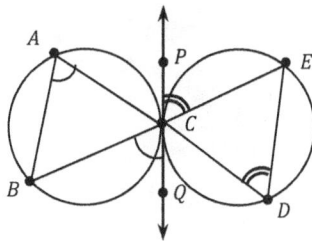

Case 1 : $\angle BCQ \cong \angle BAC$ and $\angle ECP \cong \angle EDC$

(∵ Since $\angle BCQ$ and $\angle ECP$ are vertical angles,

$\quad \angle BCQ \cong \angle ECP$.

$\quad$ Therefore, $\angle BAC \cong \angle EDC$

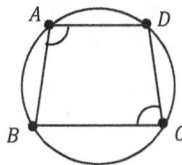

Since the alternate interior angles are congruent, $\overline{AB} \,/\!/\, \overline{DE}$.

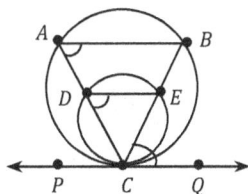

Case 2 : Since $\angle ECQ \cong \angle EDC$ and $\angle ECQ \cong \angle BAC$,

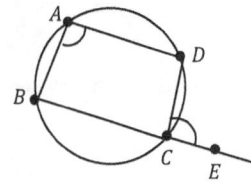

$\angle EDC \cong \angle BAC$.

Since the corresponding angles are congruent,

$\overline{AB} \,/\!/\, \overline{DE}$.

Exercises

#1   Given a circle with center P, let $\overline{CB} = \overline{PB}$. Find $m\ \widehat{CB}$; $m\ \widehat{AC}$; $m\ \widehat{ACB}$; $m\ \widehat{CAB}$.

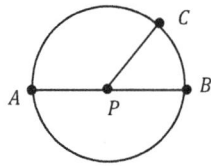

(not to scale)

#2   Given a circle with center P, let $\overline{AC} = \overline{BD}$ be diameters of the circle and $m\ \angle DBC = 30°$. Find the measure of each minor arc of the circle.

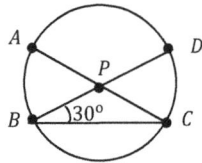

#3   The point P is the circumcenter of the triangles $\Delta\ ABC$. Find the measure of $\angle a$ for each triangle.

(1)

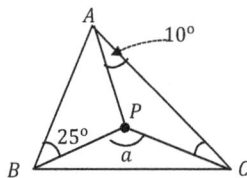

(2)

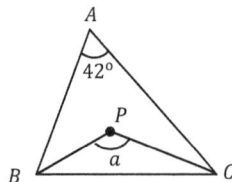

(3)

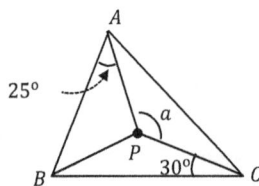

(4)

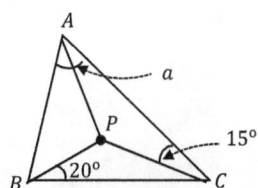

(5)

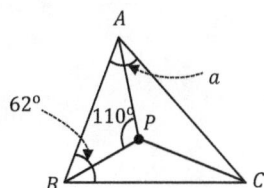

(6)

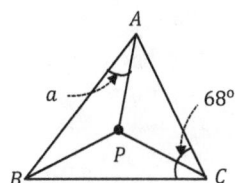

#4 The point Q is the incenter of the triangles $\triangle ABC$. Find the measure of $\angle a$ for each triangle.

(1)

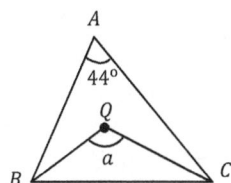

(2)

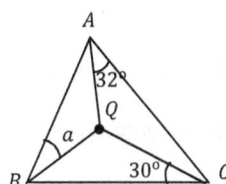

(3)

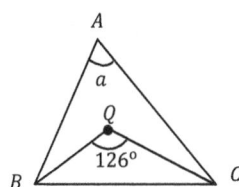

(4)

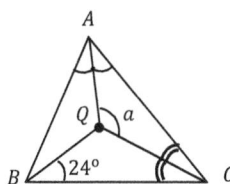

(5)

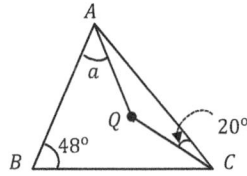

#5   For the following problems, the point Q is the incenter of the triangle $\triangle ABC$. Let $\overline{DE} \mathbin{/\mkern-5mu/} \overline{BC}$.

(1)

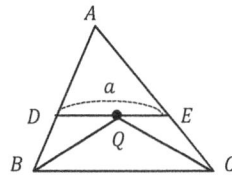

Let $AD = 4$; $DB = 2$; $AE = 5$; $EC = 3$.

Find the value of a and the perimeter of $\triangle ADE$.

(2)

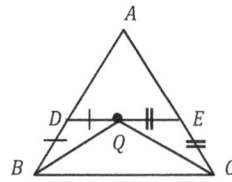

Let $AB = 10$; $AC = 8$; $BC = 6$.

Find the perimeter of $\triangle ADE$.

#6

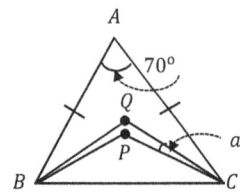

The point P and Q are the circumcenter and incenter of the triangle $\triangle ABC$, respectively.

Let $\overline{AB} \cong \overline{AC}$ and $m \angle A = 70°$. Find the measure of $\angle a$.

#7

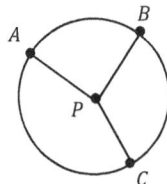

For a circle with center P,

$m \widehat{AB} : m \widehat{BC} : m \widehat{CA} = 2 : 3 : 4$.

Find the measures of $\angle APB : \angle BPC : \angle APC$.

#8

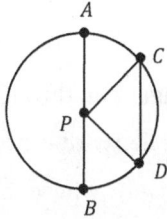

For a circle with center P, $\overline{AB} \parallel \overline{CD}$ and $m\,\overset{\frown}{BD} = \frac{1}{2}m\,\overset{\frown}{CD}$

Find the measure of $\angle CPD$.

#9

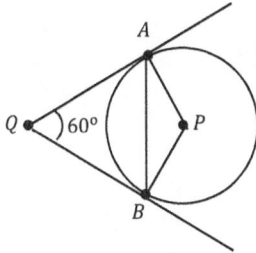

Two tangents through an exterior point Q intersect a circle (with center P) at points A and B. The measure of $\angle Q$ is $60°$. Find the measure of $\angle PBA$.

#10

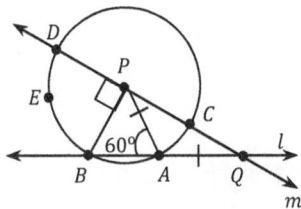

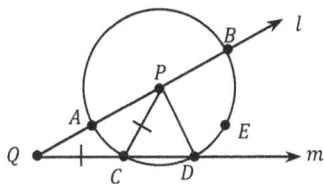

Secants l and m, through an exterior point Q, intersect a circle (with center P) at points A and B ; C and D , respectively, where $\overline{CD}$ is a diameter of the circle. Let $m\angle PAB = 60°$, $\overline{QA} \cong \overline{AP}$, and $m\,\overset{\frown}{DEB} = 3$ inches. Find $m\,\overset{\frown}{AC}$.

#11

Secants l and m, through an exterior point Q, intersect a circle (with center P) at points A and B ; C and D , respectively, where $\overline{AB}$ is a diameter of the circle.   Let $\overline{QC} \cong \overline{CP}$ and $m\,\overset{\frown}{AC} = a°$.
Find $m\,\overset{\frown}{BED}$.

#12

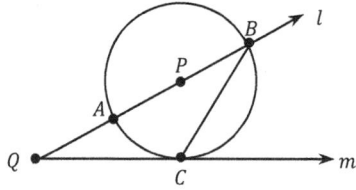

A secant l and a tangent m, through an exterior point Q, intersect a circle (with center P) at points A and B ; C, respectively, where $\overline{AB}$ is a diameter of the circle.

(1) Let $m \angle Q = 32°$. Find $m \angle QBC$.

(2) Let $m \angle APC = 70°$. Find $m \angle ACQ$.

(3) Let $m \angle QBC = 35°$. Find $m \angle ACQ$.

#13 For the following, find the value of :

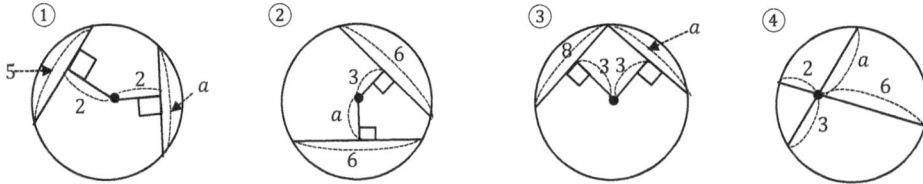

#14 For the following, find the measure of $\angle Q$:

(1)

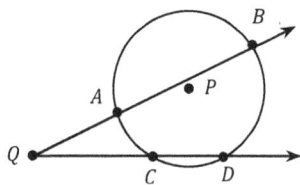

$m \, \widehat{AC} = 2$ and $m \, \widehat{BD} = 6$

(2)

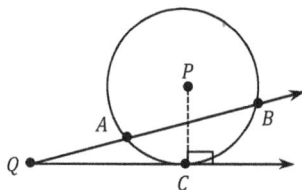

$m \, \widehat{AC} = 3$ and $m \, \widehat{BC} = 5$

(3)

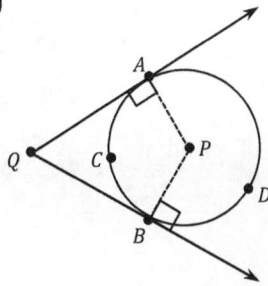

$m \, \widehat{ACB} = 4$ and $m \, \widehat{ADB} = 7$

#15   For the following, find the value of x:

(1)

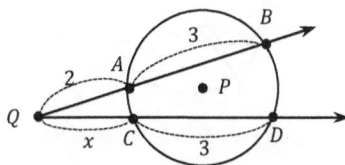

$QA = 2$ and $AB = CD = 3$

(2)

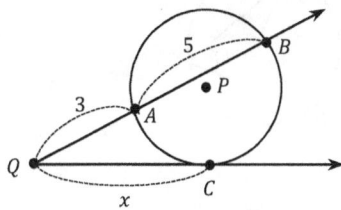

$QA = 3$ and $AB = 5$

(3)

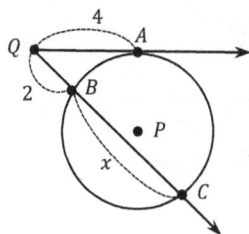

$QA = 4$ and $QB = 2$

(4)

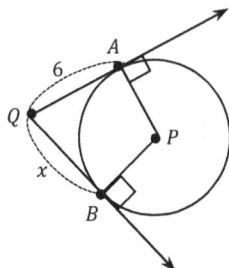

$QA = 6$

#16

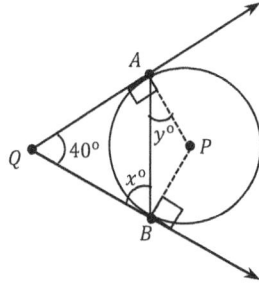

$\overline{PA} \perp \overline{AQ}$, $\overline{PB} \perp \overline{BQ}$.

Let $m \angle Q = 40°$. Find the values of x and y.

#17 For the following circles with centers P, find the measure of $\angle a$:

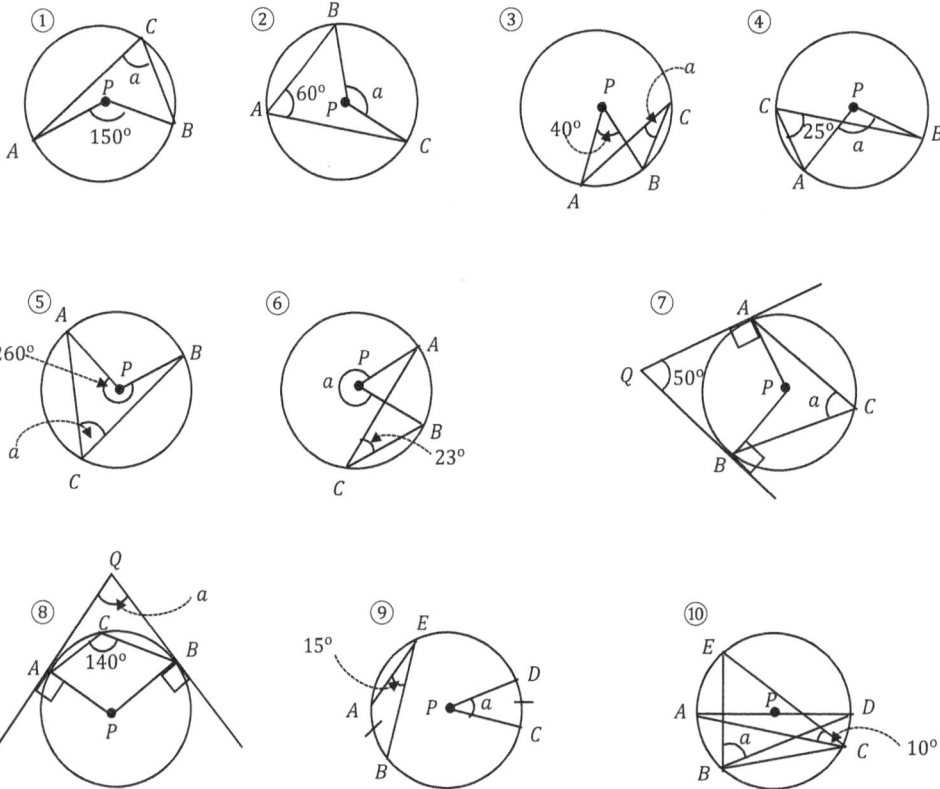

①   ②   ③   ④

⑤   ⑥   ⑦

⑧   ⑨   ⑩

#18 For the following circles (P is a center), find the measures of $\angle a$, $\angle b$, and $\angle c$:

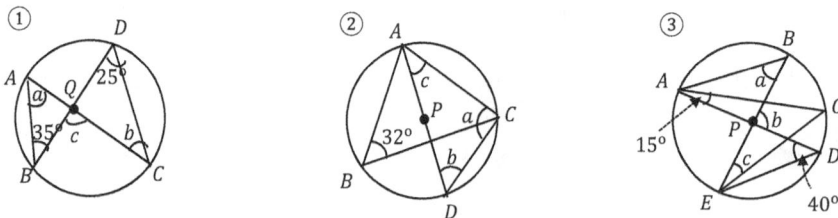

①   ②   ③

#19 For the following circumscribed quadrilaterals $\square ABCD$ about a circle, find the value of x :

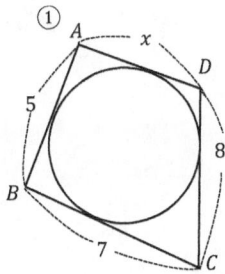

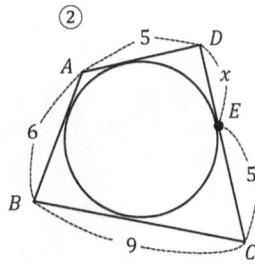

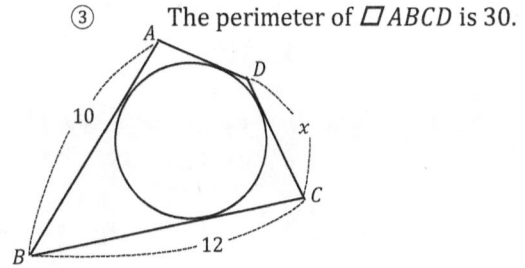

①

②

③ The perimeter of $\square ABCD$ is 30.

#20 For the following inscribed quadrilaterals $\square ABCD$ in a circle,

find the measures of $\angle a$ and $\angle b$ for ① and ③

and find the measure of $\angle a$ for ② :

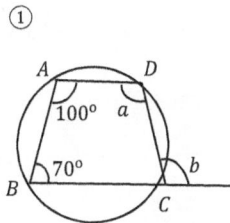

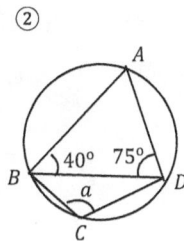

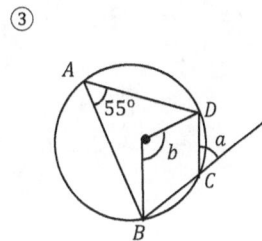

①

②

③

Geometric Constructions

CHAPTER 5

Chapter 5. Geometric Constructions

5-1 Basic Geometric Constructions

1. Concurrence

If two or more lines intersect at exactly one point, then the lines are concurrent.

(1) The Perpendicular Bisector Concurrence

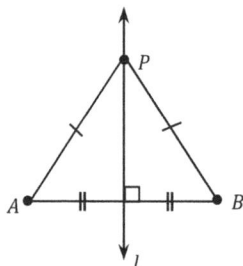

The perpendicular bisector l of a segment $\overline{AB}$ is the set of all points which are equidistant from the end points A and B of the segment $\overline{AB}$.

$PA = PB$

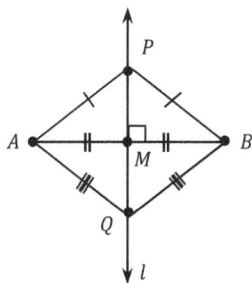

$\overline{AB} \perp \overleftrightarrow{PQ}$, $AM = BM$

$\Rightarrow PA = PB$ and $QA = QB$

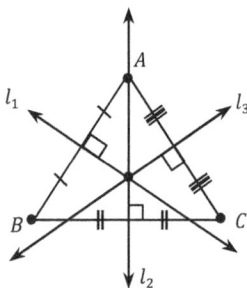

The perpendicular bisectors l_1, l_2 and l_3 of the sides $\overline{AB}$, $\overline{BC}$, and $\overline{AC}$ of a triangle $\triangle ABC$ are concurrent.

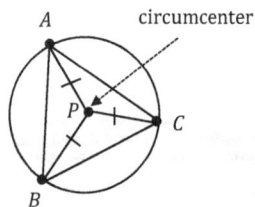

The perpendicular bisectors of the sides of a triangle intersect at a point P. The intersection point P is called the circumcenter.

So, the point P is equidistant from the vertices of the triangle. $PA = PB = PC$

(2) The Altitude Concurrence

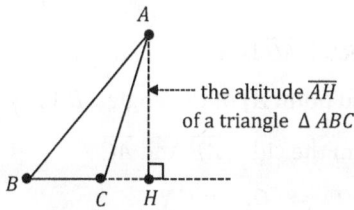

The altitude(height) of a triangle is the length of the perpendicular segment from a vertex of the triangle to the opposite side.

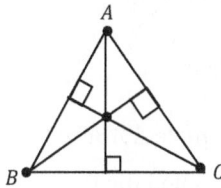

The three altitudes of a triangle $\triangle ABC$ are concurrent.

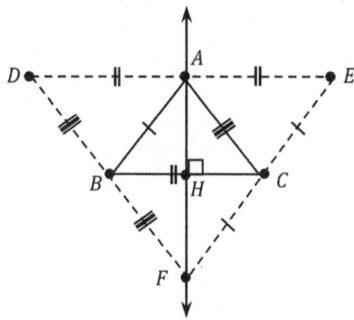

($\because$ For a triangle $\triangle ABC$, $\overline{AH}$ is the altitude through a point A to the opposite side $\overline{BC}$.

Consider the line $\overleftrightarrow{AH}$ instead of the segment $\overline{AH}$ because the altitudes are segments and the segments do not necessarily intersect at all. Through each vertex, Look at the three lines parallel to the opposite sides.

Then $\overline{DE} \parallel \overline{BC}$, $\overline{DF} \parallel \overline{AC}$, and $\overline{EF} \parallel \overline{AB}$.

Since the opposite sides of a parallelogram are congruent,

$DA = AE = BC$, $AB = CF = CE$, and $AC = DB = BF$.

Since $DA = AE = BC$, $\overleftrightarrow{AH}$ is the perpendicular bisector of $\overline{DE}$.

Similarly, the other two altitudes of $\triangle ABC$ are the perpendicular bisectors of the other sides.

Since the perpendicular bisectors of the sides of a triangle are concurrent, these altitudes are concurrent.)

(3) The Angle Bisector Concurrence

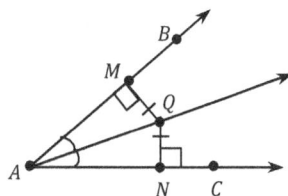

The angle bisector $\overrightarrow{AQ}$ is the set of all interior points (except the end point A) of the angle $\angle BAC$ which are equivalent from the sides $\overleftrightarrow{AB}$ and $\overleftrightarrow{AC}$.

$$\angle BAQ \cong \angle CAQ \quad \Rightarrow \quad QM = QN$$

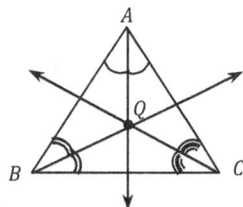

The angle bisectors of $\angle A$, $\angle B$, and $\angle C$ of a triangle $\triangle ABC$ are concurrent.

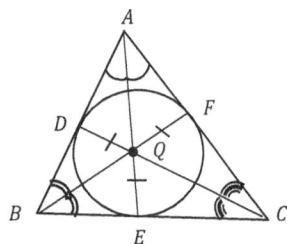

The angle bisectors of a triangle $\triangle ABC$ intersect at a point Q. The point Q is called the incenter. So, the point Q is equidistant from the sides of the triangle.

$$QD = QE = QF$$

(4) The Median Concurrence

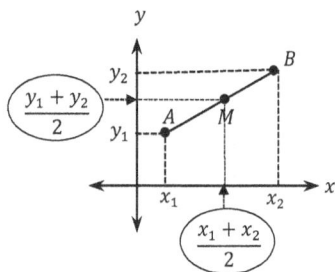

1) Midpoint formula :

   For any points $A = (x_1 , y_1)$ and $B = (x_2 , y_2)$, the midpoint of $\overline{AB}$ is the point

   $$M = \left(\frac{x_1 + x_2}{2} , \frac{y_1 + y_2}{2} \right).$$

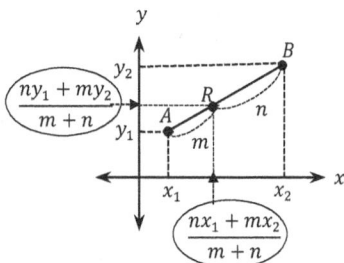

2) For any points $A = (x_1 , y_1)$ and $B = (x_2 , y_2)$, the point R is between A and B.

   If $AR : RB = m : n$, then

   $$R = \left(\frac{nx_1 + mx_2}{m+n} , \frac{ny_1 + my_2}{m+n} \right).$$

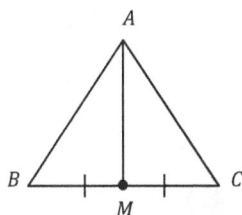

The median of a triangle is a segment that connecting a vertex and the midpoint of the opposite side.

If a point M is the midpoint of $\overline{BC}$, the opposite side of $\angle A$, then the segment $\overline{AM}$ is called the *median* from A to $\overline{BC}$.

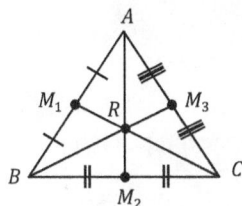

The medians of all triangle are concurrent. The point of concurrency of the medians is called the *centroid* of the triangle.

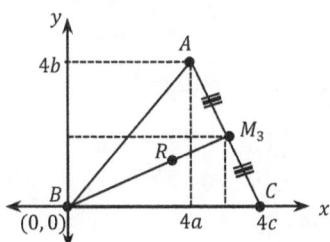

$\because$ Let $A = (4a, 4b)$, $B = (0, 0)$ and $C = (4c, 0)$.

Since M_3 is the midpoint of $\overline{AC}$,

$$M_3 = \left(\frac{4a+4c}{2}, \frac{4b+0}{2} \right) = (2a + 2c, 2b).$$

Let a point R be on the median $\overline{BM_3}$.

Suppose $BR = 2RM_3$; $BR : RM_3 = 2 : 1$

Then by the midpoint formula,

$$R = \left(\frac{0 \cdot 1 + 2 \cdot (2a+2c)}{2+1}, \frac{0 \cdot 1 + 2 \cdot (2b)}{2+1} \right) = \left(\frac{4a+4c}{3}, \frac{4b}{3} \right)$$

Since M_2 is the midpoint of $\overline{BC}$, $M_2 = (2c, 0)$.

Now, let a point R' be on the median $\overline{AM_2}$.

Suppose $AR' = 2R'M_2$; $AR' : R'M_2 = 2 : 1$

Then, by the midpoint formula,

$$R' = \left(\frac{1 \cdot 4a + 2 \cdot 2c}{2+1}, \frac{1 \cdot 4b + 2 \cdot 0}{2+1} \right) = \left(\frac{4a+4c}{3}, \frac{4b}{3} \right)$$

So, $R = R'$

Similarly, the corresponding point R'' of the median $\overline{CM_1}$ is equal to the same point R. $R = R' = R''$

Therefore, our assumption is proved.

Hence, the intersecting point R is two-thirds of each median from the vertex to the opposite side.

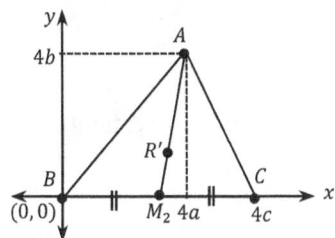

$AR : RM_2$
$= BR : RM_3$
$= CR : RM_1$
$= 2 : 1$

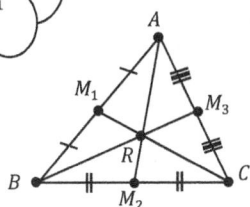

$AR = \frac{2}{3} AM_2$, $BR = \frac{2}{3} BM_3$,

$CR = \frac{2}{3} CM_1$

5-2 Basic Constructions

1. Triangle Construction

For the three lengths of the sides of a triangle $\triangle ABC$, the sum of two of the lengths must be greater than the third length, the longest one.

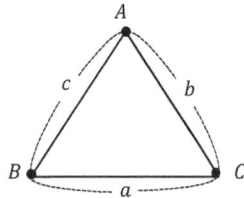

$a < b + c$, where a is the longest length.

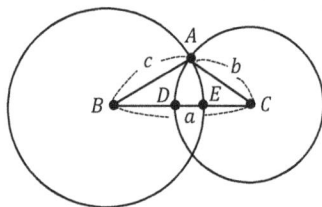

Consider two circles of radius b and c with a as the length between their centers B and C.

Since $a = b + c - DE$, $a < b + c$. (a is the longest length of the side of the triangle.)

Therefore, we can construct a triangle with three lengths of sides.

If the condition $a < b + c$ (a is the longest length.) is not satisfied, then we cannot construct a triangle.

For example,

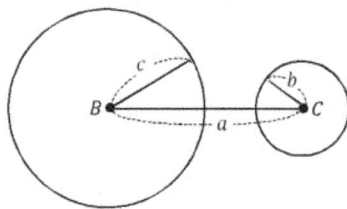

$a > b + c$

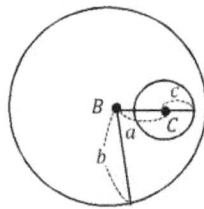

$b > a + c$

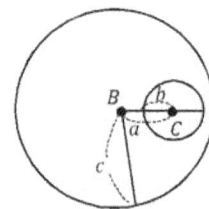

$c > a + c$

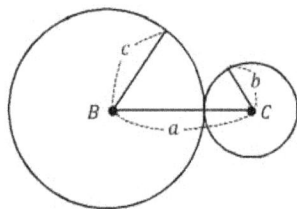

$a = b + c$

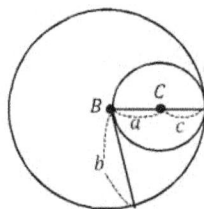

$b = a + c$

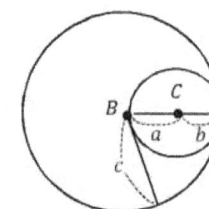

$c = a + b$

2. Angle Bisector Construction

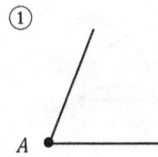

① Given $\angle A$.

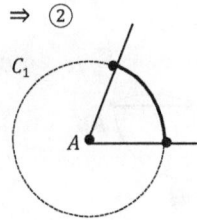

⇒ ② A is the center of a circle C_1.

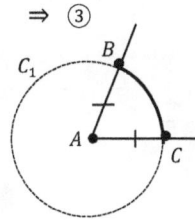

⇒ ③ The circle and the sides of $\angle A$ intersect at points B and C.

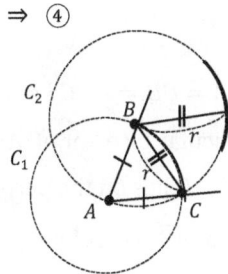

⇒ ④ B is the center of a circle C_2 with radius $r = BC$

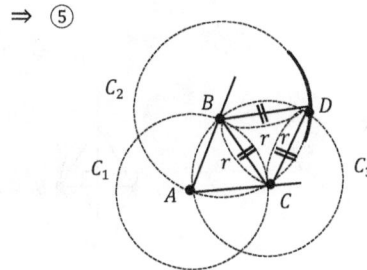

⇒ ⑤ C is the center of a circle C_3 with radius $r = BC$

⇒ ⑥ Two circles C_2 and C_3 Intersect at points A and D lying on opposite sides of $\overline{BC}$

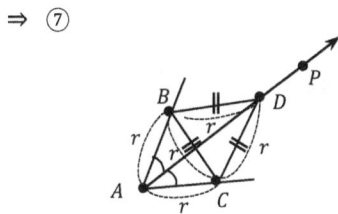

⇒ ⑦ By *SSS* Postulate, $\triangle ABD \cong \triangle ACD$ Therefore, $\angle BAD \cong \angle CAD$ and $\overrightarrow{AP}$ is the bisector.

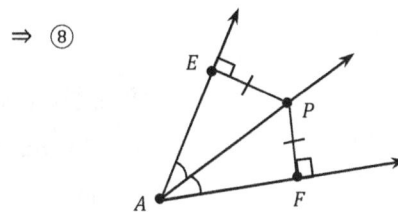

⇒ ⑧ $\angle EAP \cong \angle FAP$ ⇒ $PE = PF$

3. Perpendicular Bisector Construction

①   ⇒ ②   ⇒ ③

Begin with segment $\overline{AB}$.

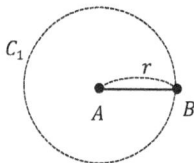

A is the center of circle C_1 with radius $r = AB$.

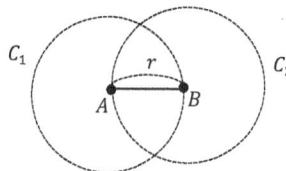

B is the center of circle C_2 with radius $r = AB$.

⇒ ④   ⇒ ⑤

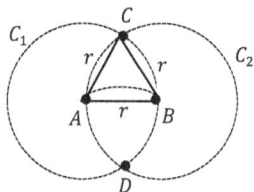

The circles intersect at two points, C and D.

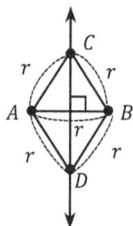

Since $CA = CB = r$, C lies on the perpendicular bisector of $\overline{AB}$. Similarly, D lies on the perpendicular bisector of $\overline{AB}$. Therefore, $\overleftrightarrow{CD}$ is the perpendicular bisector of $\overline{AB}$.

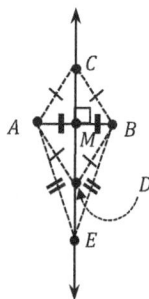

⇒ ⑥ $\overline{AB} \perp \overleftrightarrow{CD}$ and $AM = MB$
⇒ $AC = BC = AD = BD$,
and $AE = BE$.

4. Perpendicular Construction

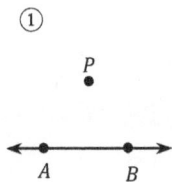

①   ⇒ ②   ⇒ ③

P is an external point for a given line $\overleftrightarrow{AB}$.

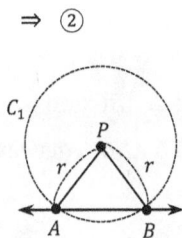

P is the center of circle C_1 with radius $r = PA = PB$.
The line $\overleftrightarrow{AB}$ and the circle C_1 intersect at two points, A and B.

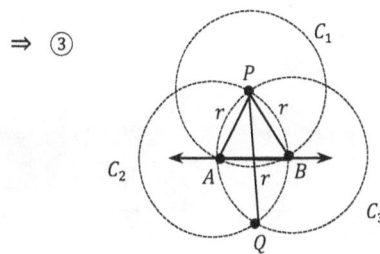

A is the center of circle C_2 with radius $r = PA$.
B is the center of circle C_3 with radius $r = PB$.

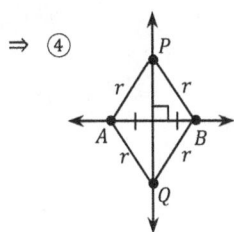

⇒ ④

Construct the perpendicular bisector of $\overline{AB}$.

Since $AP = BP = r$, the perpendicular bisector passes through the point P. Similarly, we get $AQ = BQ = r$.

Since $\overleftrightarrow{PQ}$ is the line containing points P and Q, which satisfies $PA = PB$ and $QA = QB$,

$\overleftrightarrow{PQ}$ is the perpendicular bisector of $\overline{AB}$.

Exercises.

#1

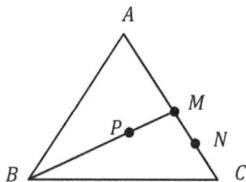

For a triangle $\triangle ABC$, $\overline{BM}$ is a median of $\triangle ABC$ and the point P is the centroid of $\triangle ABC$. Answer the following .

(1) $PM = 2$. Find the length of $\overline{BM}$.

(2) $BP = 8$. Find the length of $\overline{PM}$ and $\overline{BM}$.

(3) $A = (2, 6)$, $C = (6, 4)$. Find the coordinates of M.

(4) A point N is on $\overline{AC}$ and $AN : NC = 3 : 1$. Find the coordinates of N.

#2 Given following three lengths of the sides of a triangle, determine whether we can construct a triangle.

(1) $4, 5, 6$

(2) $3, 5, 9$

(3) $8, 10, 20$

(4) $2, 2, 2$

(5) $3, 3, 6$

#3

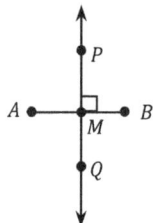

The line $\overleftrightarrow{PQ}$ is the perpendicular bisector of a segment $\overline{AB}$. Determine whether the following expressions are true or false.

(1) $AP = BP$ (2) $AB = PQ$ (3) $AM = MB$

(4) $AP = AQ$ (5) $m\angle AMQ = m\angle PMB$ (6) $AQ = BQ$

#4

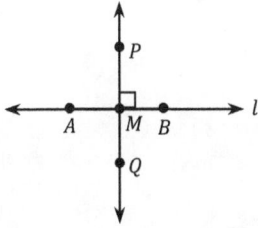

For a line *l* and a point *P,* the point *P* is an external point and the point *Q* is on the line $\overleftrightarrow{PQ}$. $\overleftrightarrow{PQ}$ is perpendicular to a given *l,* through the point *P.* Determine whether the following expressions are true or false.

(1) $AP = BP$ (2) $AQ = BQ$ (3) $AP = AQ$

(4) $AM = BM$ (5) $m \angle PMA = m \angle PMB$ (6) $PM = MQ$

#5

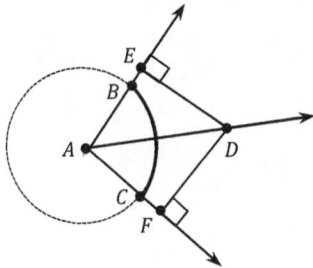

The point *D* in the interior of $\angle BAC$ is on the bisector of $\angle BAC$. Determine whether the following expressions are true or false.

(1) $AB = AC$ (2) $m \angle BAD = m \angle CAD$ (3) $BD = CD$

(4) $AB = BD$ (5) $\overline{AB} \perp \overline{BD}$ (6) $ED = DF$

Polygonal Regions and their Areas

Chapter 6 Polygonal Regions and their Areas

6-1 Polygonal Regions

1. **Polygonal Regions**
   (1) **Definition**
   (2) **Areas of Squares and Rectangles**
   (3) **Perimeters of Squares and Rectangles**
2. **Triangles and Quadrilaterals**
   (1) **Areas of Triangles**
   (2) **Parallelograms and Rhombi**
       1) **Parallelograms**
       2) **Rhombi**
   (3) **Trapezoids**
   (4) **Regular Polygons**

6-2 Circles

1. **The Circumference C of a Circle**
   (1) **The number π**
   (2) $C = 2\pi r$
2. **The Area A of a Circle ; $A = \pi r^2$**

6-3 Arcs and Sectors

1. **The Length L of an Arc ; $L = 2\pi r \dfrac{x}{360}$**
2. **The Area S of a sector ; $S = \dfrac{1}{2} rL = \pi r^2 \dfrac{x}{360}$**

CHAPTER 6

Chapter 6. Polygonal Regions and their Areas

6-1 Polygonal Regions

1. Polygonal Regions

(1) Definition

1)   A triangular region is the union of a triangle and its interior.

2)   A Polygonal region is the union of a finite number of triangular regions.

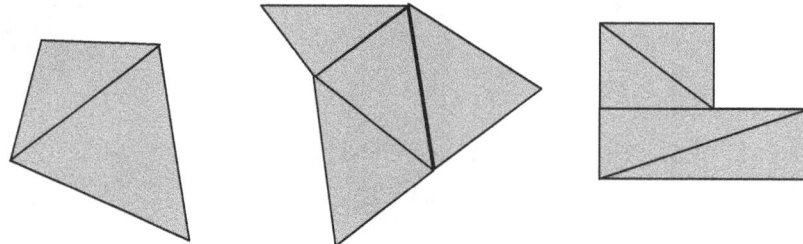

(2) Areas of Squares and Rectangles

The area of a square is a^2.

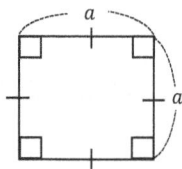

The unit postulate :

The area of a square region is defined as the square (a^2) of the length (a) of its edge.

> The area of a rectangle is $A = bh$.

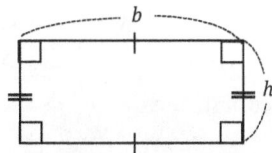

The area of a rectangle is defined as the product of the length of its base and the length of its altitude.

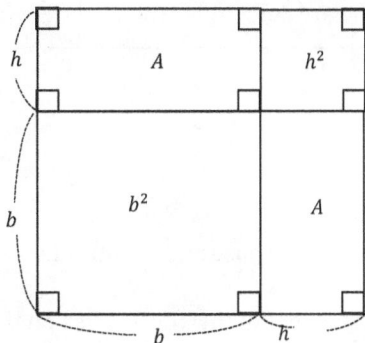

($\because$ Since the areas of two squares are b^2 and h^2 by the unit postulate and the total polygon is a square with a length $b + h$,

the area of the total polygon is $(b + h)^2$.

So, $b^2 + h^2 + 2A = (b + h)^2$.

Since $(b + h)^2 = (b + h)(b + h) = b^2 + 2bh + h^2$,

$2A = 2bh$.

Therefore, $A = bh$.)

> The perimeter is the sum of the lengths of all the sides.

> Area and perimeter include units.
> Perimeter shares the same units as the length of sides.
> But area includes the squaring unit.

(3) Perimeters of Squares and Rectangles :

Since all four sides of a square are congruent, the perimeter of a square with a length a is $4a$.

Since a rectangle is a parallelogram with four right angles, opposite sides are congruent.

So, the perimeter of a rectangle with base b and altitude h is $2b + 2h$.

2. Triangles and Quadrilaterals

(1) Areas of Triangles

> The area of $\triangle ABC$ is
> $\frac{1}{2} \times$ base $\times$ altitude
> $= \frac{1}{2} bh$

1) The area of a right triangle is equal to half the product of its legs.

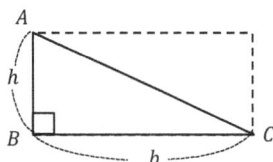

> The area of a right triangle is $\frac{1}{2} bh$.

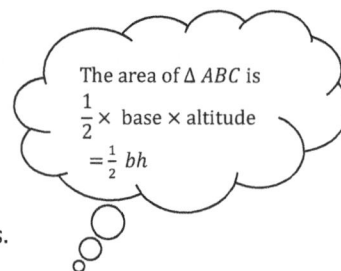

2) The area of a triangle is equal to half the product of its base and its altitude.

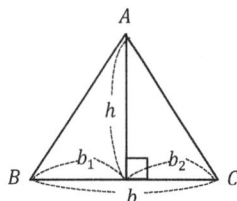

Case 1: If the foot of the altitude lies on the base, then consider two triangles with bases b_1 and b_2 , where $b = b_1 + b_2$.

Then, the area of $\triangle ABC$ is the sum of the areas of these two triangles.

The area of $\triangle ABC$ is

$$\frac{1}{2} b_1 h + \frac{1}{2} b_2 h = \frac{1}{2}(b_1 + b_2)h = \frac{1}{2} bh .$$

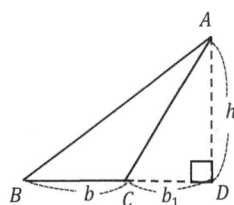

Case 2: If the foot of the altitude does not lie on the base, then consider two right triangles with bases b_1 and $b + b_1$.

Then, the area of right triangle $\triangle ABD$ is the sum of the areas of $\triangle ABC$ and $\triangle ACD$.

$$\frac{1}{2}(b + b_1)h = \text{Area of } \triangle ABC + \frac{1}{2} b_1 h$$

So, the area of $\triangle ABC$ is

$$\frac{1}{2}(b + b_1)h - \frac{1}{2} b_1 h = \frac{1}{2} bh + \frac{1}{2} b_1 h - \frac{1}{2} b_1 h = \frac{1}{2} bh.$$

The area of $\triangle ABC$ is the area of $\triangle DEF$.

3) If two triangles have the same base b and the same altitude h, then they have the same area because the area of each triangle is $\frac{1}{2} bh$.

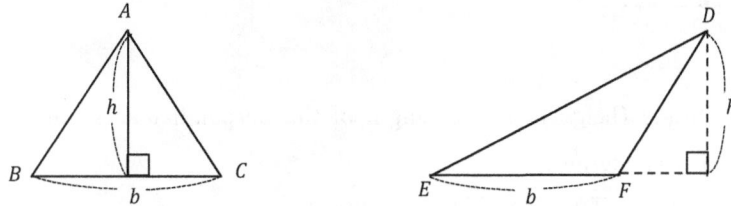

4) If two triangles have the same altitude h but different bases, then the ratio of their areas is equal to the ratio of their bases.

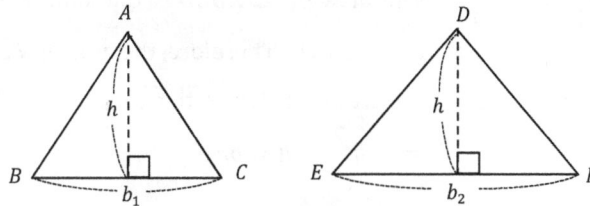

($\because$ Since the area of $\triangle ABC$ is $\frac{1}{2} b_1 h$ and the area of $\triangle DEF$ is $\frac{1}{2} b_2 h$,

the area of $\triangle ABC$: the area of $\triangle DEF$ = $\frac{1}{2} b_1 h$: $\frac{1}{2} b_2 h$ = b_1 : b_2)

(2) Parallelograms and Rhombi

1) Parallelograms

The area of a parallelogram is the product of its base and its altitude.

Two opposite sides of a parallelogram are congruent.

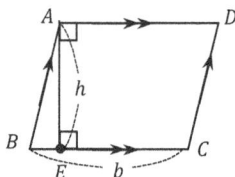

> The area of a parallelogram is bh.

($\because$ The altitude (height) h is the length of the perpendicular segment connecting opposite sides of the parallelogram.

Since $\overline{AE} \perp \overline{BC}$ and $\overline{AE} \perp \overline{AD}$, $\overline{AE}$ is the altitude.

Also, $\overline{BC}$ and $\overline{AD}$ are bases of the parallelogram $\square ABCD$.

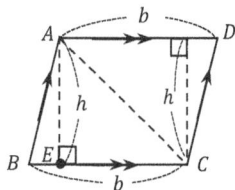

Since either diagonal divides the parallelogram into two triangles, consider the triangles $\triangle ABC$ and $\triangle ACD$.

The area of $\square ABCD$ is the sum of the area of $\triangle ABC$ and $\triangle ACD$. Therefore, the area of $\square ABCD$ is

the area of $\triangle ABC$ + the area of $\triangle ACD$

$= \frac{1}{2}bh + \frac{1}{2}bh = bh$.)

2) Rhombi

The area of a rhombus is the product of half the product of its diagonals.

> The area of a rhombus is $\frac{1}{2}d_1 d_2$.

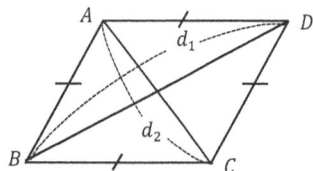

($\because$ The diagonals of a rhombus are perpendicular to each other. Since a rhombus is a parallelogram, the diagonals bisect each other.

Consider two triangles $\triangle ABD$ and $\triangle CBD$.

Then the area of $\square ABCD$

= the area of $\triangle ABD$ + the area of $\triangle CBD$

$= \frac{1}{2}d_1 h + \frac{1}{2}d_1 h = d_1 h$.

Since $h = \frac{1}{2}d_2$, the area of rhombus $\square ABCD$ is

$d_1 h = d_1 \left(\frac{1}{2}d_2 \right) = \frac{1}{2}d_1 d_2$.)

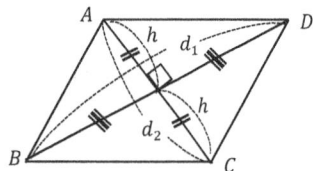

(3) Trapezoid

The area of a trapezoid is half the product of its altitude and the sum of its bases.

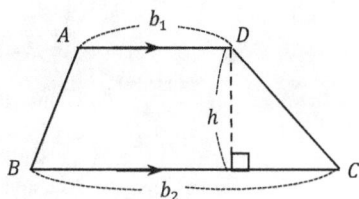

> The area of a trapezoid is $\frac{1}{2}h(b_1 + b_2)$.

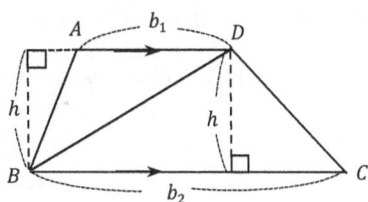

(∵ Either diagonal divides the trapezoid into two triangles.

Consider the triangles $\triangle ABD$ and $\triangle BCD$.

Then the area of $\square ABCD$

= the area of $\triangle ABD$ + the area of $\triangle BCD$

$= \frac{1}{2}b_1h + \frac{1}{2}b_2h = \frac{1}{2}h(b_1 + b_2)$.)

(4) Regular Polygons

If a polygon (n-gon, $n \geq 3$) is

① convex,

② all of its sides are congruent, and

③ all of its angles are congruent,

then the polygon is called a *regular polygon*.

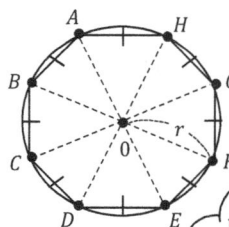

Inscribed regular octagon (8-gon) in a circle with center 0 and radius r.

Consider a circle with center 0 and radius r.

Dividing the circle into n congruent arcs, the measure of each of the arcs is equal to $\frac{360^\circ}{n}$.

By connecting two end points on each arc, we obtain a polygon (n-gon).

Clearly, the polygon is convex.

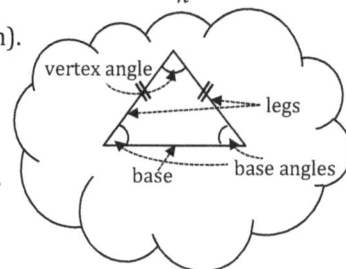

Since the arcs are congruent, the corresponding chords are congruent.

Therefore, all the sides of the polygon are congruent.

Considering the radii (from the center 0 of the circle to the end points on each arc) as sides of triangles, we obtain the number of n isosceles triangles.

By SSS Postulate, all n triangles are congruent.

So, the measure of each angle of the polygon is equal to twice the measure of each base angle of each isosceles triangle.

Therefore, all the angles of the polygon are congruent.

Hence, the polygon is regular.

The altitude (height) of the isosceles triangle in the regular polygon is called the *apothem a* of the polygon.

The *perimeter* of a regular polygon (n-gon) is

$$P = n \cdot \text{(the length of the side)} = n \cdot b$$

The *area* of a regular polygon (n-gon) is

$$A = n \cdot \text{(the area of one triangle)} = n \cdot \frac{1}{2}ab = \frac{1}{2}aP$$

: half the product of the apothom and the perimeter.

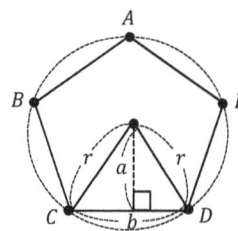

6-2 Circles

1. The Circumference C of a Circle

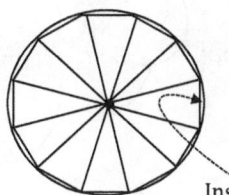

As the number of sides n of the inscribed regular polygon (n-gon) in a circle approaches infinity (∞), the perimeter P_n (as a function of n) approaches a certain fixed value, which is considered as the circumference (C) of the circle.
$$\lim_{n \to \infty} P_n = C$$

Inscribed regular polygon (n-gon) in a circle.

If the number of sides of a polygon is large enough, then the perimeter of n-gon is approximately equal to the circumference of the circle. So, we define the *circumference* of a circle as the limit of the perimeter of the inscribed regular polygon.

(1) The number π

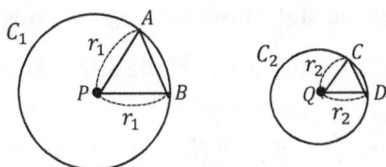

Every circle has the same ratio of the circumference to the diameter, regardless of its size.

Consider circles C_1 and C_2

where C_1 is a circle with center P and radius r_1, and C_2 is a circle with center Q and radius r_2.

Since $\dfrac{r_1}{r_2} = \dfrac{r_1}{r_2}$ (sides of the triangles), $\triangle ABP$ and $\triangle CDQ$ are proportional.

Suppose two regular polygons (n-gons) are inscribed in the two circles.

Then, the measures of the central angles of $\triangle ABP$ and $\triangle CDQ$ are the same as $\dfrac{360^\circ}{n}$.

By SAS Similarity Theorem, $\triangle ABP$ and $\triangle CDQ$ are similar.

See Chapter 8, 8-2 for more information.

That is,

$$\frac{AB}{r_1} = \frac{CD}{r_2} \quad ; \quad \frac{n \cdot AB}{r_1} = \frac{n \cdot CD}{r_2} \quad ; \quad \frac{\text{the perimeter of } n\text{-gon in } C_1}{r_1} = \frac{\text{the perimeter of } n\text{-gon in } C_2}{r_2}.$$

Since the circumferences of C_1 and C_2 are the limits of the perimeters of the two polygons,

$$\frac{\text{the circumference of } C_1}{r_1} = \frac{\text{the circumference of } C_2}{r_2} \quad ; \quad \frac{\text{the circumference of } C_1}{2\,r_1} = \frac{\text{the circumference of } C_2}{2\,r_2}$$

Therefore,

$$\frac{\text{the circumference of } C_1}{\text{the diameter of } C_1} = \frac{\text{the circumference of } C_2}{\text{the diameter of } C_2} \quad : \text{ same ratio of the circumference to the diameter.}$$

The ratio of the circumference of a circle to the diameter of the circle is denoted by π.

(2) $C = 2\pi r$

Let C be the circumference of a circle with radius r.

Then, $\pi = \dfrac{\text{the circumference of a circle}}{\text{the diameter of the circle}} = \dfrac{C}{2\,r}$.

Therefore, the circumference of a circle with radius r is $C = 2\pi r$.

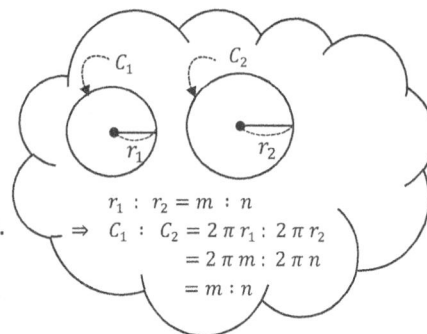

$$r_1 : r_2 = m : n$$
$$\Rightarrow C_1 : C_2 = 2\pi r_1 : 2\pi r_2$$
$$= 2\pi m : 2\pi n$$
$$= m : n$$

The number π is not a rational number.

However, we use the approximately closed rational number 3.14 or $\dfrac{22}{7}$ to write the value numerically for the number .

2. The Area A of a Circle ; $A = \pi r^2$

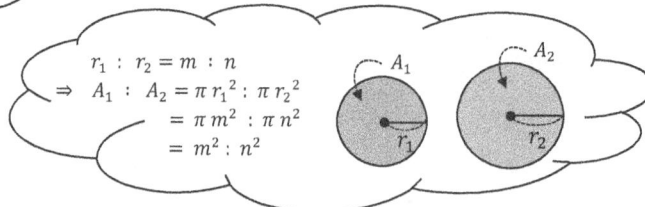

If the number of sides of a polygon inscribed in a circle is large enough, then the area of the polygon (n -gon) is approximately equal to the area of the circle.

So, we define the area A of a circle as the limit of the area of the inscribed regular polygon

denote, $\dfrac{1}{2}aP \longrightarrow A$ where a is the apothem and P is the perimeter of the inscribed regular polygon.

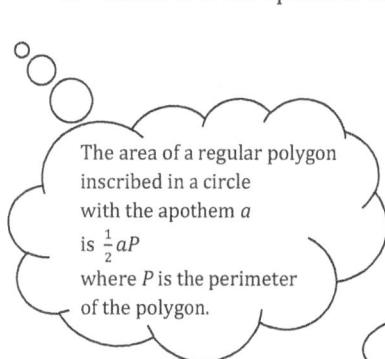

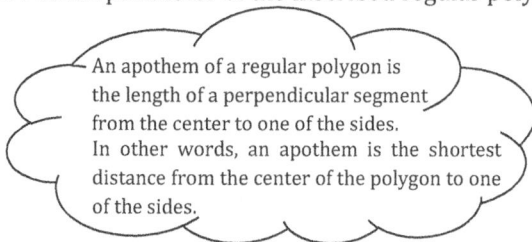

> The area of a regular polygon inscribed in a circle with the apothem a is $\frac{1}{2}aP$ where P is the perimeter of the polygon.

> An apothem of a regular polygon is the length of a perpendicular segment from the center to one of the sides. In other words, an apothem is the shortest distance from the center of the polygon to one of the sides.

$$r_1 : r_2 = m : n$$
$$\Rightarrow A_1 : A_2 = \pi r_1^2 : \pi r_2^2$$
$$= \pi m^2 : \pi n^2$$
$$= m^2 : n^2$$

$\boxed{A = \pi r^2}$

If the number of sides of a polygon inscribed in a circle is large enough, then the distance between the apothem and the radius is small enough. So, the apothem a is appoximately equal to the radius r of the circle ; denote, $a \longrightarrow r$.

Since the circumference C of a circle is the limit of the perimeter P of the inscribed regular polygon

denote, $P \longrightarrow C$, we can get $\dfrac{1}{2}aP \longrightarrow \dfrac{1}{2}rC$.

Since $\dfrac{1}{2}aP \longrightarrow A$, $A = \dfrac{1}{2}rC$.

Since $C = 2\pi r$, the area of a circle with radius r is $A = \dfrac{1}{2}rC = \dfrac{1}{2}r(2\pi r) = \pi r^2$.

6-3 Arcs and Sectors

1. The Length L of an Arc ; $L = 2\pi r \dfrac{x}{360}$

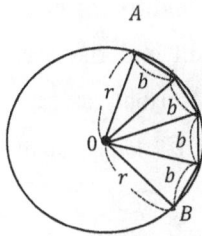

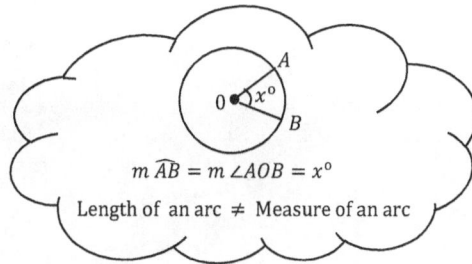

$$m \, \widehat{AB} = m \, \angle AOB = x^\circ$$

Length of an arc $\neq$ Measure of an arc

Since each chord has the same length b, the perimeter P of aregular polygon (n-gon) is $P = n \cdot b$.

If n is large enough, then the length of an arc $\widehat{AB}$ is approximately equal to $n \cdot b$.

So, we define the length of arc $\widehat{AB}$ as the limit of P.

Consider a circle as an arc of measure 360°. Then the circumference of the circle can be considered the length of an arc of measure 360°.

$$\boxed{L = 2\pi r \dfrac{x}{360}}$$

Two circles with the same radius have the same ratio of the lengths of two arcs to the measures of their arcs.

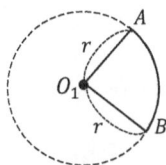

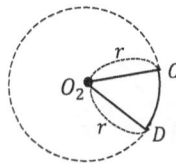

$$\frac{\text{The length of arc } \widehat{AB}}{m \, \widehat{AB}} = \frac{\text{The length of arc } \widehat{CD}}{m \, \widehat{CD}}$$

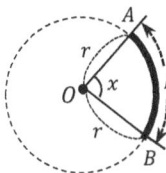

Consider a circle with radius r.

If $m \, \widehat{AB} = x$, then the length of the arc $\widehat{AB}$ is

$$L = 2\pi r \frac{x}{360}.$$

($\because$ Since $\dfrac{\text{The length of arc } \widehat{AB}}{m \, \widehat{AB}} = \dfrac{\text{The circumference of a circle with radius } r}{360}$, $\dfrac{L}{x} = \dfrac{2\pi r}{360}$.

Therefore, $L = \dfrac{2\pi r \cdot x}{360} = 2\pi r \cdot \dfrac{x}{360}$)

2. The Area S of a sector ; $S = \frac{1}{2}rL = \pi r^2 \frac{x}{360}$

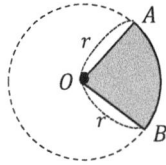

A sector is a part of a circular region, the union of a circle and its interior, bounded by two radii and the arc. Also, a sector is described as the union of all segments $\overline{OP}$ for any point P on an arc $\overparen{AB}$.

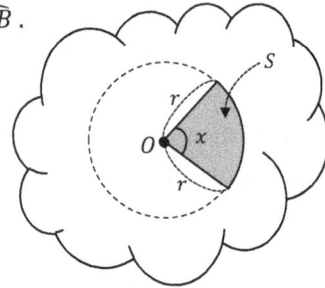

$$S = \frac{1}{2}rL = \pi r^2 \frac{x}{360}$$

($\because$ Since the circumference C of a circle with radius r is $C = 2\pi r$ and the area A of a circle with radius r

is $A = \pi r^2$, $A = \pi r^2 = \pi r \left(\frac{C}{2\pi}\right) = \frac{1}{2}rC$.

For a sector with arc length L and area S, $\quad \frac{L}{S} = \frac{C}{A} = \frac{C}{\frac{1}{2}rC} = \frac{2}{r}$; $2S = rL$; $S = \frac{1}{2}rL$

Since $L = 2\pi r \frac{x}{360}$, $\quad S = \frac{1}{2}rL = \frac{1}{2}r\left(2\pi r \frac{x}{360}\right) = \pi r^2 \frac{x}{360}$.)

Exercises

#1 Find the area of the polygons $\square ABCD$.

   (1) The diagonals of rhombus $\square ABCD$ intersect at point P.

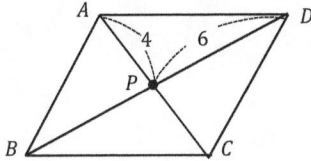

   (2) The trapezoid $\square ABCD$ has an altitude of 6.

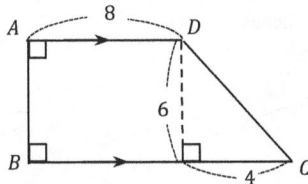

#2 A regular polygon has a perimeter of 60 and an apothem of 8. Find the area of the polygon.

#3 The area of a regular octagon with an apothem of 4 is 160.
     Find the length of each side of the octagon.

#4 Find the perimeter and area for the following :

   (1)

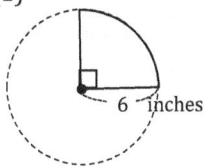

   (2)

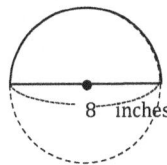

#5 For a circle with a circumference of 14 and the measure of arc $m\,\widehat{AB} = 30$,
     answer the following.

   (1) Find the area of the circle.

   (2) Find the length of $\widehat{AB}$.

   (3) Find the area of the sector with arc $\widehat{AB}$.

#6 Find the length of the arc $\widehat{AB}$ and the area of the shaded part.

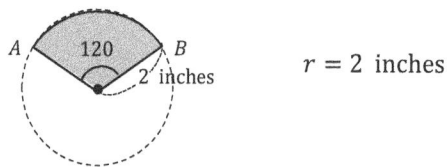

$r = 2$ inches

#7 Find the value of x for the following :

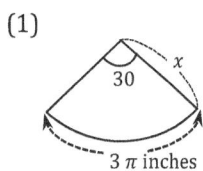

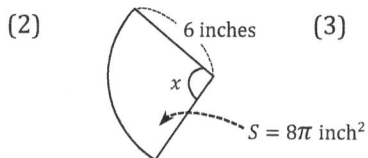

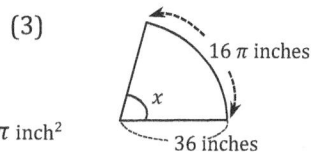

(1)

$3\,\pi$ inches

(2)   6 inches

$S = 8\pi$ inch^2

(3) $16\,\pi$ inches

36 inches

#8 Find the area of the shaded part in each figure :

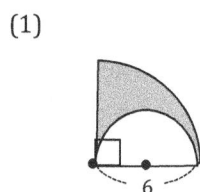

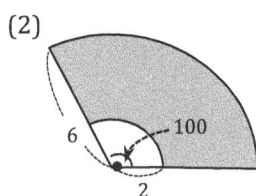

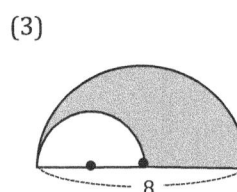

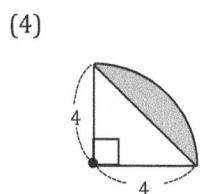

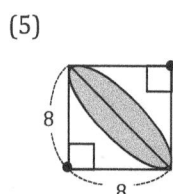

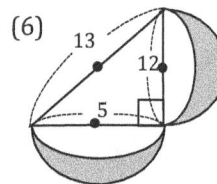

(1)

6

(2)

6   100

2

(3)

8

(4)

4

4

(5)

8

8

(6)   13

12

5

#9 For two circles with radii r_1, r_2, circumferences C_1, C_2 and areas A_1, A_2, respectively,

the ratio of r_1 to r_2 is $2:3$. Answer the following.

(1) Find the ratio of the circumferences of the circles.

(2) Find the ratio of the areas of the circles.

#10 $\overset{\frown}{AB} : \overset{\frown}{BC} : \overset{\frown}{AC} = 2 : 3 : 4$

Find the measures of $\angle AOB$ $\angle BOC$, and $\angle AOC$.

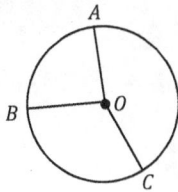

#11 Find the ratio of $\overset{\frown}{OB} : \overset{\frown}{AB}$.

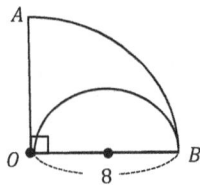

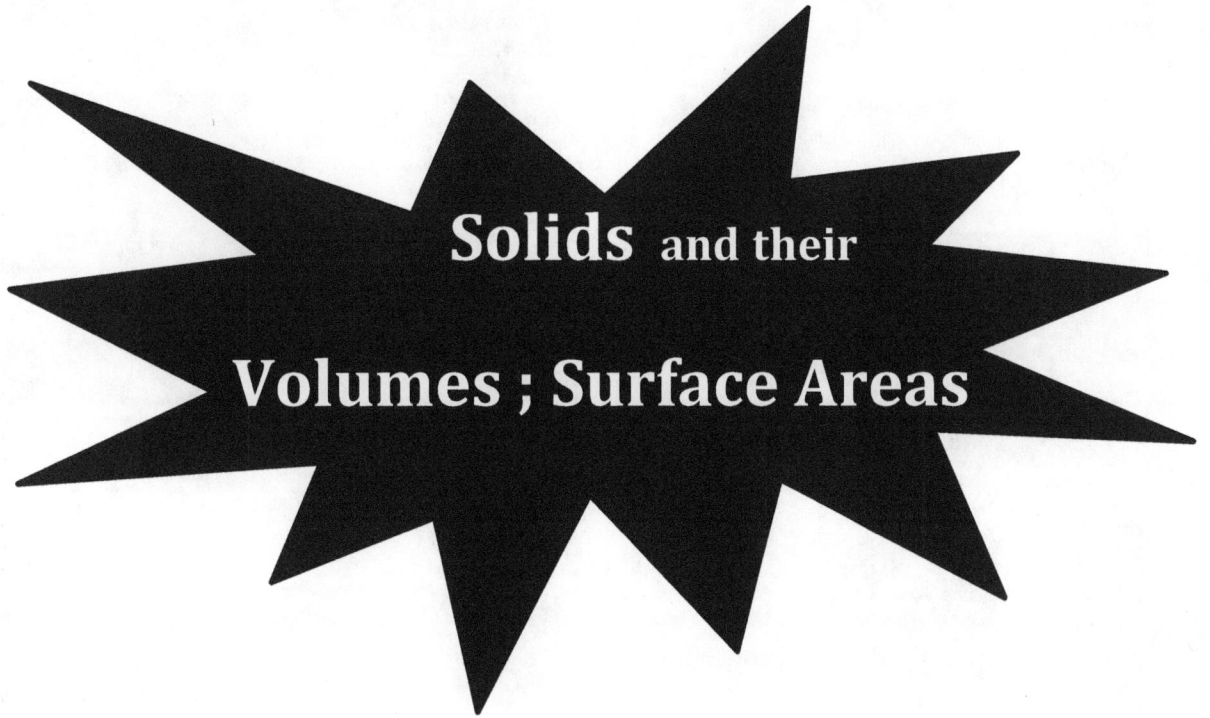

Solids and their

Volumes ; Surface Areas

Chapter 7   Solids and their Volumes; Surface Areas

7-1   Prisms

1.   Definition
2.   Classification of Prisms
3.   Volume and Surface Area
    (1) Volume of Prisms
    (2) Surface Area of Prisms

7-2   Pyramids

1.   Definition
2.   Classification of Pyramid
3.   Volume and Surface Area
    (1) Volume of Pyramids
    (2) Surface Area of Pyramids

7-3   Cylinders and Cones

1.   Cylinders
    (1) Definition
    (2) Volume and Surface Area
2.   Cones
    (1) Definition
    (3) Volume and Surface Area

7-4   Spheres

CHAPTER
7

Chapter 7. Solids and their Volumes ; Surface Areas

7-1 Prisms

1. Definition

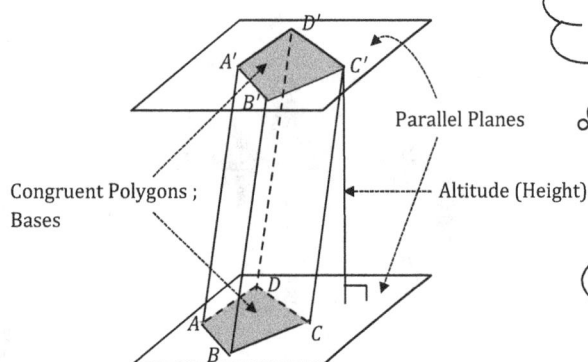

$\square\,B'BCC'$; lateral face

$\overline{BB'}$, $\overline{CC'}$; lateral edges

The lateral faces of a prism are parallelograms.

The lateral faces of a right prism are rectangles.

A *prism* consists of two congruent polygons (called *bases*) in two parallel planes connected by parallelograms, called *lateral faces*.

If the lateral edges of the prism are perpendicular to the bases, then the prism is called a *right prism*.

If the prism is not a right, then it is called an *oblique prism*.

If the congruent polygons of a right prism are regular, then the right prism is called a *regular prism*.

The union of lateral faces and the two bases of a prism is called the *surface* of a prism.

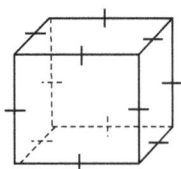

For example, a *cube* is a rectangular prism with 2 square bases and 4 lateral faces. All of the edges of a cube are congruent.

2. Classification of Prisms (Prisms are described by their bases.)

Prism	Base of Prism	Number of Lateral Faces, f	Number of Vertices, v	Number of Lateral Edges, e	$f + v - e = 2$
Triangular Prism	3-gon	5	6	9	2
Rectangular Prism or Cube	4-gon	6	8	12	2
Pentagonal Prism	5-gon	7	10	15	2
Hexagonal Prism	6-gon	8	12	18	2
⋮	⋮	⋮	⋮	⋮	⋮
⋮	n-gon	$n+2$	$2n$	$3n$	2

3. Volume and Surface Area

(1) Volume of Prisms

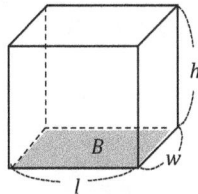

The volume of a rectangular prism is

$$V = B \cdot h \quad , \quad$$ where B is the area of a base and h is the altitude (height)

$$= l \cdot w \cdot h$$

Height = b Length= a

Height = c

$$V = \frac{1}{2}abh$$ $$V = \frac{1}{2}abh + \frac{1}{2}ach$$

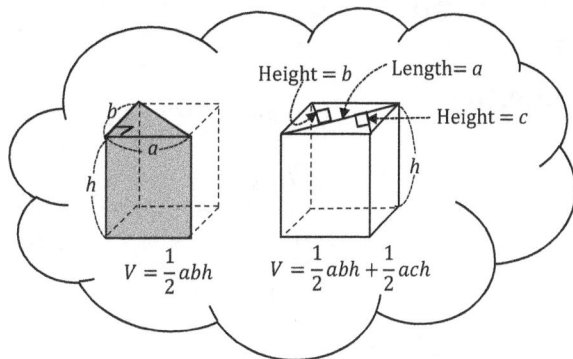

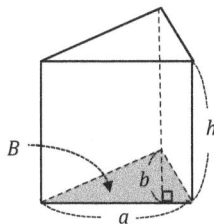

The volume of a triangular prism is

$$V = B \cdot h \quad , \quad$$ where B is the area of a base and h is the altitude (height)

$$= \left(\frac{1}{2}ab\right) \cdot h = \frac{1}{2}abh$$

(2) Surface Area of Prisms

The surface area of a prism is the sum of the areas of all lateral surfaces.

$S = 2 \times$ (area of base) $+$ $\underset{\parallel}{\underline{\text{all areas of lateral faces}}}$
(perimeter of base) $\times$ (height)

That is,

the surface area of a prism

= Areas of bases + All areas of lateral faces

= $2 \cdot$ (Area of base) + (Perimeter of base $\cdot$ height)

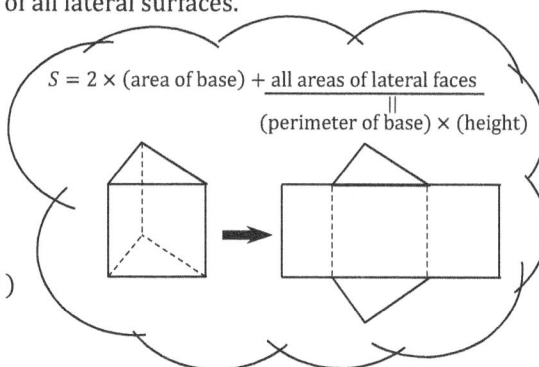

The surface area of a rectangular prism is $S = 2lh + 2wh + 2lw$.

The surface area of a triangular prism is $S = 2\left(\frac{1}{2}ab\right) + 3(ah) = ab + 3ah = a\,(b + 3h)$.

Example

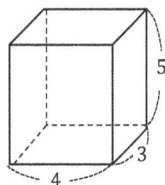

The volume of the prism is

$V = B \cdot h = (4 \cdot 3) \cdot 5 = 60$ cubic units.

The surface area of the prism is

$S = 2 \cdot (4 \cdot 3) + (4 + 3 + 4 + 3) \cdot 5$

$= 24 + 70 = 94$ square units.

7-2 Pyramids

1. Definition

Vertex

Polygon ; Base

Plane

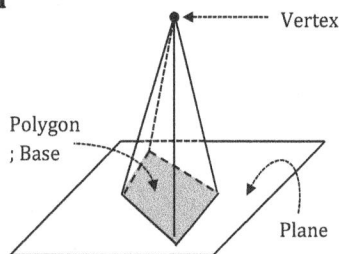

A *pyramid* consists of a polygon (called a *base*) in a plane and triangular lateral faces with a point called a *vertex* which is not in the plane.

The lateral edges of a pyramid connect the vertices of the bases with the vertex of the pyramid.

If the base of the pyramid is a regular polygon, then the pyramid is called a *regular pyramid.*

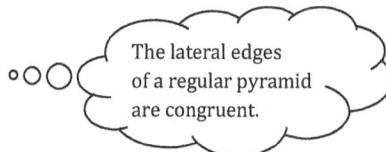

The lateral edges of a regular pyramid are congruent.

The altitude (height) of a pyramid is the perpendicular distance from the vertex to the base plane.

The altitude of a regular pyramid is called the *axis* which is perpendicular to the base at its center.

The altitude of one of the triangular lateral faces of a regular pyramid is called the *slant height*.

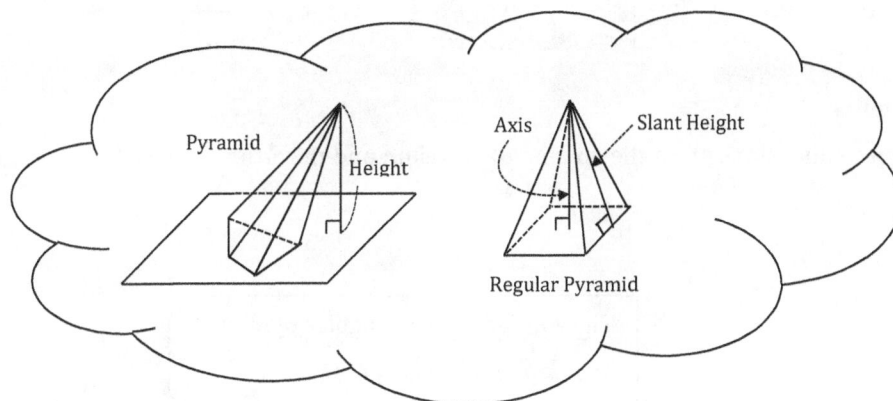

2. Classification of Pyramids (Pyramids are described by their bases.)

Pyramid	Base of Pyramid	Number of Lateral Faces	Number of Vertices	Number of Lateral Edges
Triangular Pyramid	3-gon	4	4	6
Rectangular Pyramid or Cube	4-gon	5	5	8
Pentagonal Pyramid	5-gon	6	6	10
Hexagonal Pyramid	6-gon	7	7	12
$\vdots$	$\vdots$	$\vdots$	$\vdots$	$\vdots$
$\vdots$	n-gon	$n+1$	$n+1$	$2n$

If we pour the water from the pyramid into a cube, then the cube will become $\frac{1}{3}$ of the way full.

$\frac{1}{3}$ of cube

3. Volume and Surface Area

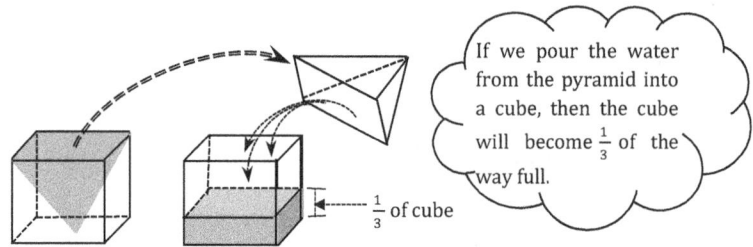

(1) Volume of Pyramids

The volume of a pyramid is one-third the volume of a prism ; one-third the product of its base area and its altitude.

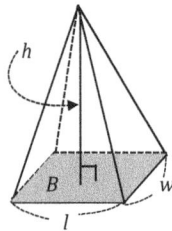

> The volume of a rectangular pyramid
> $$V = \frac{1}{3}B \cdot h = \frac{1}{3}l \cdot w \cdot h$$

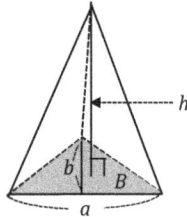

> The volume of a triangular pyramid
> $$V = \frac{1}{3}B \cdot h = \frac{1}{3} \cdot \left(\frac{1}{2}ab\right) \cdot h = \frac{1}{6}abh$$

(2) Surface Area of Pyramids

The surface area of a pyramid is the sum of its base area and the area of its lateral faces.

That is, $S = $ (area of base) + (area of lateral faces)

$\quad = $ (area of base) + ($\frac{1}{2}$ · perimeter of base · slant height)

Slant height

Slant height

One base and four lateral faces

Example

Altitude (height) = 4, Slant height = 5

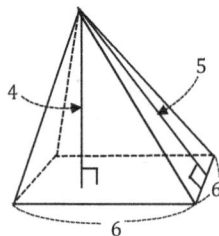

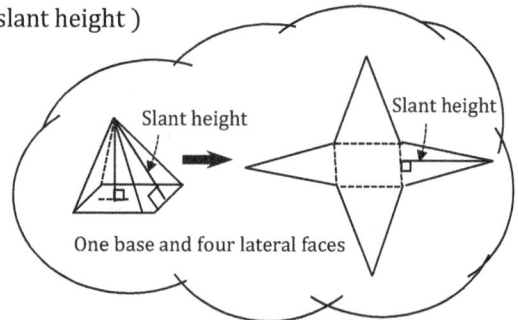

The volume of the pyramid is

$$V = \frac{1}{3}B \cdot h = \frac{1}{3}6 \cdot 6 \cdot 4 = 48 \text{ cubic units.}$$

The surface area of the pyramid is

$$S = (6 \cdot 6) + \left(\frac{1}{2} \cdot (6 + 6 + 6 + 6) \cdot 5\right)$$
$$= 36 + 60 = 96 \text{ square units.}$$

7-3   Cylinders and Cones

1. Cylinders

> Right cylinders and right prisms are exactly the same except the shape of the bases.

(1) Definition

A *cylinder* consists of two congruent circles (called bases) in two parallel planes and a lateral surface.

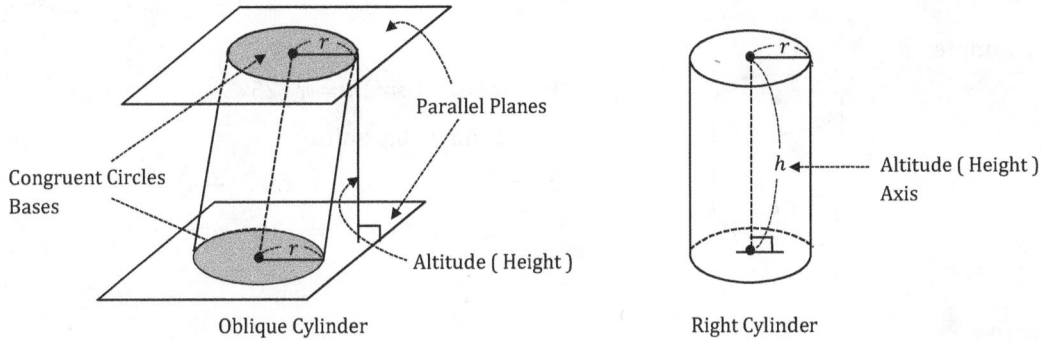

Oblique Cylinder   Right Cylinder

If the altitude (height) of the cylinder is the perpendicular segment connecting the centers of the two bases, then the cylinder is called a *right cylinder* and if not, it is called an *oblique cylinder*.

(2) Volume and Surface Area

The volume of a cylinder is the product of the area of its base and its altitude.

The volume of a cylinder is

$$V = B \cdot h = \pi r^2 h \text{ , where } B \text{ is the area of base, the circle with radius } r.$$

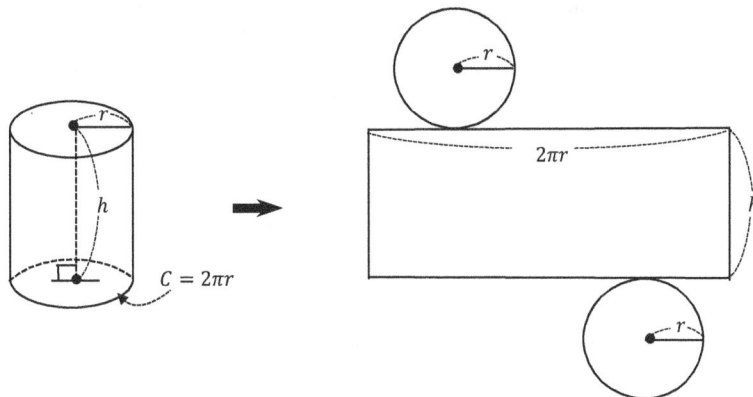

Since the lateral face of a cylinder is a rectangle, and the length of the rectangle is the circumference of the base, the surface area of a cylinder is the sum of the areas of its bases and the area of its lateral face. That is,

$$S = (\text{Areas of bases}) + (\text{Area of lateral face})$$
$$= 2(\pi r^2) + 2\pi rh = 2\pi r(r + h)$$

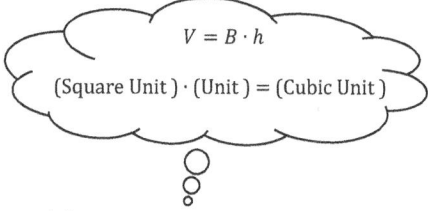

$$V = B \cdot h$$
$$(\text{Square Unit}) \cdot (\text{Unit}) = (\text{Cubic Unit})$$

Example

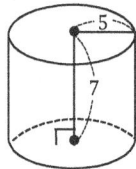

$$V = B \cdot h = (\pi r^2)h = \pi \cdot 25 \cdot 7$$
$$= 175\pi \text{ cubic units}$$
$$S = 2(\pi r^2) + 2\pi rh = 2\pi r(r + h)$$
$$= 2\pi \cdot 5 \cdot (5 + 7) = 120\pi \text{ square unit}$$

2. Cones

(1) Definition

A *cone* is exactly the same as a pyramid except for the shape of the base.

Whereas the base of a pyramid is a polygon, the base of a cone is circle.

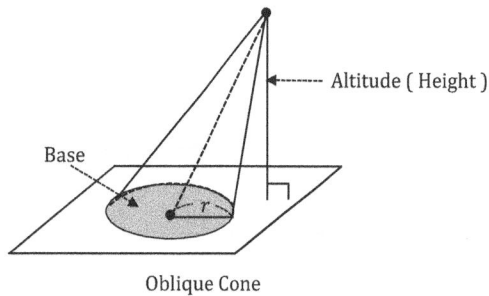

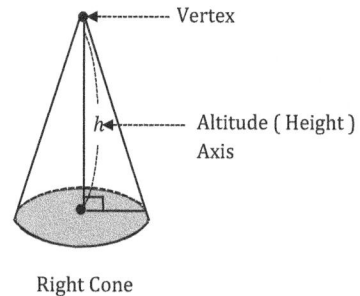

Oblique Cone

Right Cone

(2) Volume and Surface Area

The volume of a cone is one-third the product of the area of its base and its altitude.

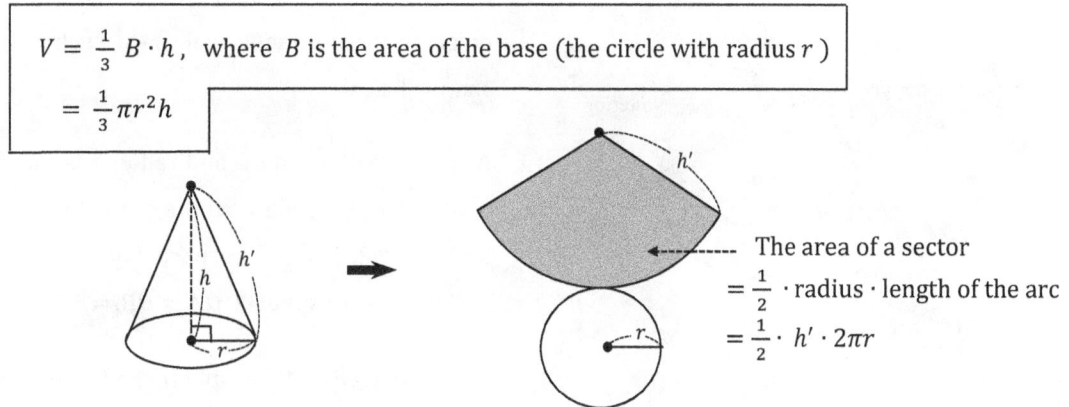

$$V = \frac{1}{3} B \cdot h, \quad \text{where } B \text{ is the area of the base (the circle with radius } r)$$
$$= \frac{1}{3} \pi r^2 h$$

The area of a sector
$= \frac{1}{2} \cdot \text{radius} \cdot \text{length of the arc}$
$= \frac{1}{2} \cdot h' \cdot 2\pi r$

The surface area of a cone is the sum of its base and the area of the lateral face.

That is, $S = (\text{area of base}) + (\frac{1}{2} \cdot \text{perimeter of base} \cdot \text{slant height})$

$= (\text{area of base}) + (\frac{1}{2} \cdot \text{circumference of base} \cdot \text{slant height})$

$= \pi r^2 + \frac{1}{2}(2\pi r \cdot h') = \pi r^2 + \pi r h' = \pi r(r + h')$

Therefore, $\boxed{S = \pi r(r + h')}$

Example

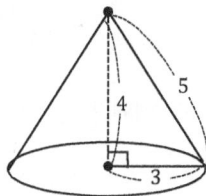

The volume of a cone with radius 3 is

$$V = \frac{1}{3} B \cdot h = \frac{1}{3}\pi r^2 h = \frac{1}{3}\pi \cdot 9 \cdot 4$$

$= 12\pi$ cubic unit.

The surface area of a cone with radius 3 and slant height 5 is

$$S = \pi r^2 + \frac{1}{2}(2\pi r \cdot h') = \pi r(r + h')$$

$= \pi \cdot 3 (3 + 5) = 24\pi$ square unit.

7-4 Spheres

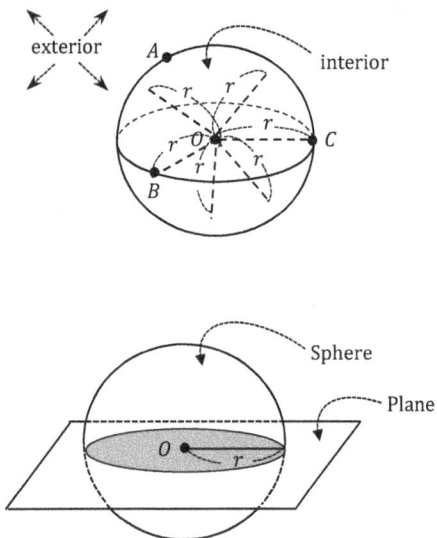

A *sphere* is a perfectly round geometrical object in three-dimensional space, such as the shape of a ball.

A sphere with center O and radius r is the set of all points which are all the same distance r from a given point O in space. For any points A, B, and C on a sphere, $OA = OB = OC = r$.

The intersection of a sphere and a plane is a circle. The largest intersection circle passes through the center of the sphere.

The intersection is called the *great circle* of the sphere.

The radius of sphere
= The radius of great circle
= r

The volume of a sphere with radius r is four-thirds the product of π and the cube of the radius.

That is,

$$V = \frac{4}{3}\pi r^3$$

The volume of a sphere is

$\frac{2}{3} \cdot$ volume of a cylinder

$= \frac{2}{3} \cdot ($ area of base $\cdot$ height $)$

$= \frac{2}{3} \cdot (\pi r^2 \cdot 2r) = \frac{4}{3}\pi r^3$, r is the radius of a sphere

The surface area of a sphere with radius r is

$$A = 4\pi r^2$$

The surface area of a sphere is

$4 \cdot$ the area of the great circle

$= 4 \cdot \pi r^2$

$= 4\pi r^2$, r is the radius of a sphere

Exercises

#1 Find the volume for the following :

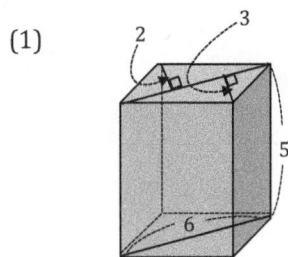

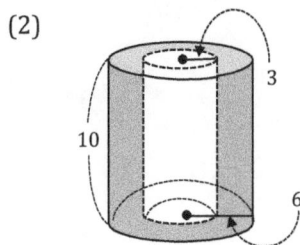

(1)

(2)

(3) The surface area of the regular prism is 294 in.^{2}

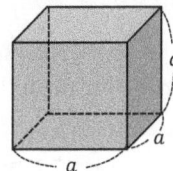

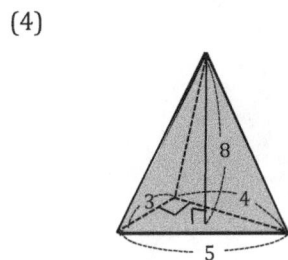

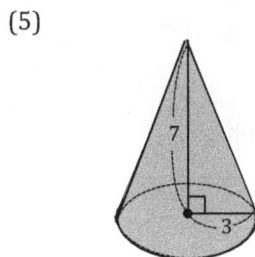

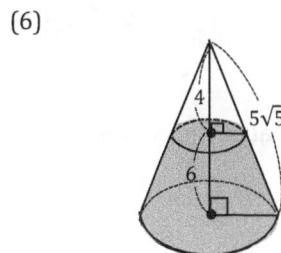

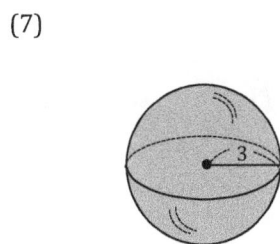

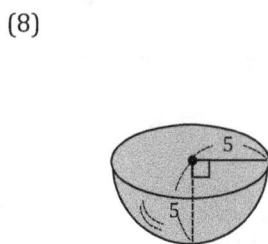

(4)

(5)

(6)

(7)

(8)

#2 Find the surface area for the following :

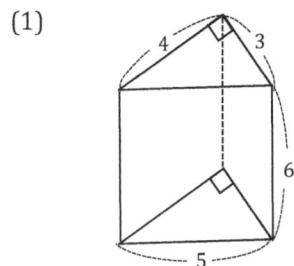

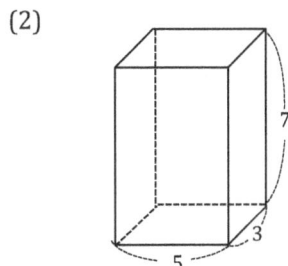

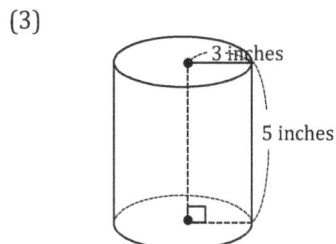

(1)

(2)

(3)

(4)

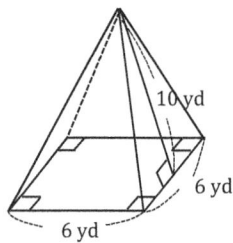

10 yd

6 yd

6 yd

(5)

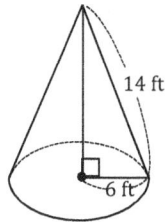

14 ft

6 ft

(6)

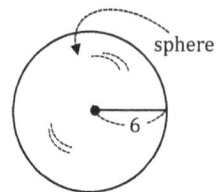

sphere

6

#3 Find the ratio for the following :

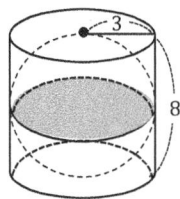

3

8

(1) Volume of the cylinder : Volume of the sphere

(2) Surface area of the cylinder : Surface area of the sphere

(3) Surface area of the sphere : Area of the great circle of the sphere

Similarity

Chapter 8   Similarity

8-1 Similarity

1. **Definition**
2. **Properties of Similarity**

8-2 Similar Triangles

1. **The Similarity Theorems**
    (1) The AAA Similarity Theorem
    (2) The AA Similarity theorem
    (3) The SSS Similarity Theorem
    (4) The SAS Similarity Theorem
2. **Similarities in Right Triangles**

8-3 Parallel Lines

1. **Triangles and Parallels**
    (1) The Proportionality Theorem and its Converse
        1) The Basic Proportionality Theorem
        2) The Converse of the Basic Proportionality Theorem
    (2) The Angle Bisector Theorems
        1) Bisector of an Interior Angle
        2) Bisector of an Exterior Angle
2. **Parallels and Transversals**

8-4 Midpoints

1. **Similar Triangles**

8-5 Centroid

1. **Similar Triangles**

8-6 Areas and Volumes

1. **Perimeter and Area**
2. **Surface Area and Volume**

Chapter 8. Similarity

8-1 Similarity

1. Definition

In general, two similar geometric figures have the same shapes, but not the same size.

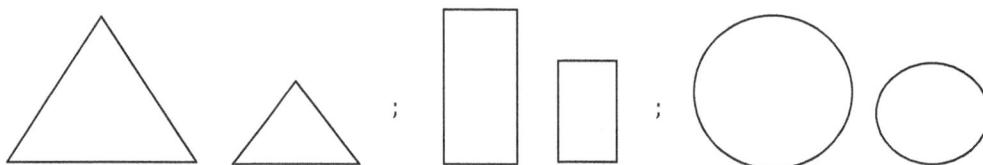

(1) Similar Trianles

For any two corresponding triangles $\triangle ABC$ and $\triangle DEF$,

if ① corresponding angles are congruent and

②   the lengths of the corresponding sides are in proportion,

then the correspondence is called a *similarity* ; $\triangle ABC \leftrightarrow \triangle DEF$, and

the triangles are *similar* ; $\triangle ABC \sim \triangle DEF$.

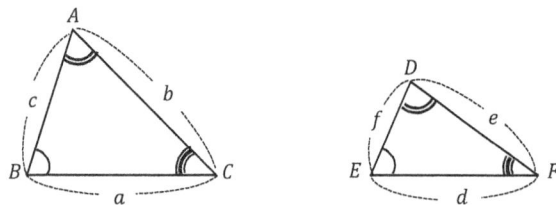

$$\left(\begin{array}{l} ① \quad \angle A \cong \angle D , \ \angle B \cong \angle E , \ \angle C \cong \angle F \\ ② \quad \dfrac{a}{d} = \dfrac{b}{e} = \dfrac{c}{f} \quad ; \ BC : EF = AC : DF = AB : DE \end{array} \right) \Rightarrow \ \triangle ABC \sim \triangle DEF$$

If $\triangle ABC \sim \triangle A'B'C'$,
then $\triangle ABC$ is an enlarged or
reduced version of $\triangle A'B'C'$
(the same shape in a different sizes).

(2) Similar Polygons

For any two corresponding solids,

if ① the ratio of the lengths of their corresponding lateral edges is a constant and

② the corresponding lateral faces are similar,

then the solids are *similar*.

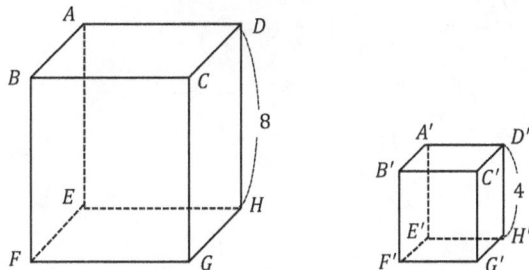

$$\left(\begin{array}{l} ① \quad \dfrac{D'H'}{DH} = \dfrac{4}{8} = \dfrac{1}{2} \; ; \; \text{constant} \\ ② \quad \square \; ABCD \sim \square \; A'B'C'D' \\ \quad\; \square \; BCFG \sim \square \; B'C'F'G' \\ \quad\; \square \; CDHG \sim \square \; C'D'H'G' \end{array}\right) \Rightarrow \text{Two prisms are similar.}$$

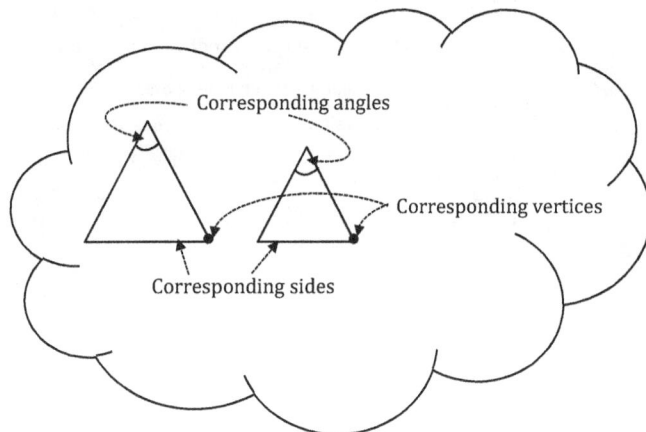

Corresponding angles

Corresponding vertices

Corresponding sides

If the two solids are similar, then their bases are similar and corresponding lengths are proportional.

2. Properties of Similarity

Consider three lines passing through the corresponding vertices of two similar triangles.
The intersection point of three lines is called the *center of similarity*, and in this case, the two triangles are directly similar to one another, their angles have the same rotational sense. The size of each triangle is proportional to its distance from the center of similarity.

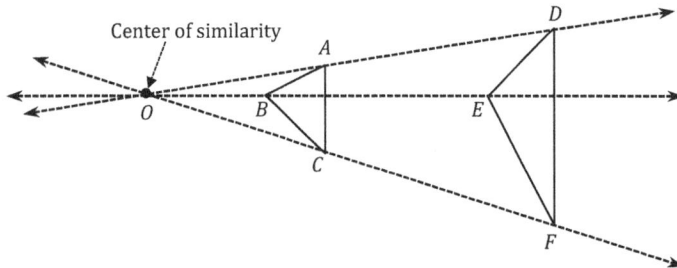

(1) The corresponding sides of the two similar triangles are parallel.
    That is, $\overline{AB} \parallel \overline{DE}$, $\overline{BC} \parallel \overline{EF}$, and $\overline{AC} \parallel \overline{DF}$

(2) The ratio of the lengths of their corresponding vertices from the center of similarity is constant and it is equal to the ratio of the lengths of their corresponding sides.
    That is, $\dfrac{OA}{OD} = \dfrac{OB}{OE} = \dfrac{OC}{OF} = \dfrac{AB}{DE}$

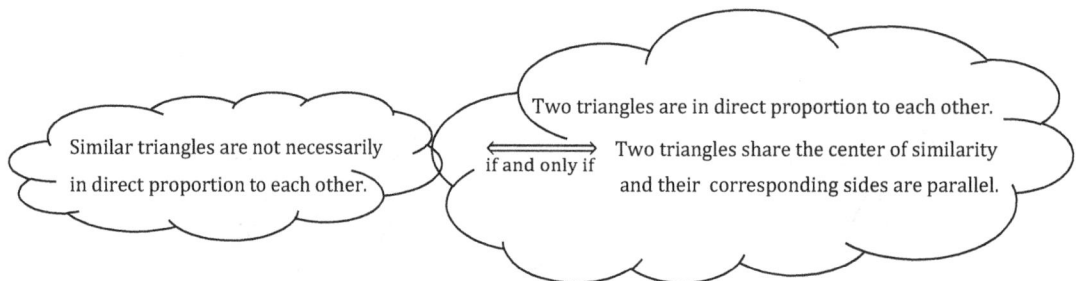

Similar triangles are not necessarily in direct proportion to each other.

if and only if

Two triangles are in direct proportion to each other.
Two triangles share the center of similarity and their corresponding sides are parallel.

8-2 Similar Triangles

> Two triangles are similar if they have the same shape, or one has the same shape as the mirror image of the other, but can be different sizes.

1. The Similarity Theorems

For any two corresponding triangles $\triangle ABC$ and $\triangle DEF$,

$\triangle ABC$ and $\triangle DEF$ are *similar*, $\triangle ABC \sim \triangle DEF$, if they satisfy the following theorems :

(1) The AAA (Angle-Angle-Angle) Similarity Theorem

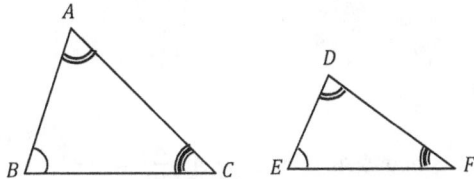

If $\angle A \cong \angle D$, $\angle B \cong \angle E$, and $\angle C \cong \angle F$,

then $\triangle ABC \sim \triangle DEF$.

Three identical angles

(2) The AA (Angle-Angle) Similarity Theorem

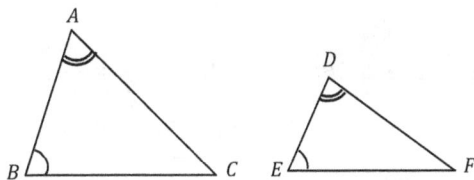

If two pairs of corresponding angles are congruent

that is,   if $\angle A \cong \angle D$, $\angle B \cong \angle E$ or

$\angle B \cong \angle E$, $\angle C \cong \angle F$ or

$\angle A \cong \angle D$, $\angle C \cong \angle F$, then

$\triangle ABC \sim \triangle DEF$.

Two identical angles

(3) The SSS (Side-Side-Side) Similarity Theorem

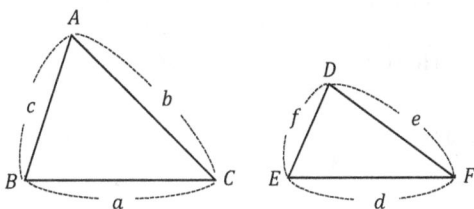

If corresponding sides are proportional

that is,   if $\dfrac{a}{d} = \dfrac{b}{e} = \dfrac{c}{f}$, then

$\triangle ABC \sim \triangle DEF$.

(4) The SAS (Side-Angle-Side) Similarity Theorem

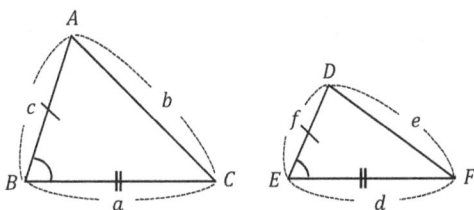

If two pairs of corresponding sides are proportional, and the included angles are congruent. That is, if $\dfrac{c}{f} = \dfrac{a}{d}$ and $\angle B \cong \angle E$, then $\triangle ABC \sim \triangle DEF$.

2. Similarities in Right Triangles

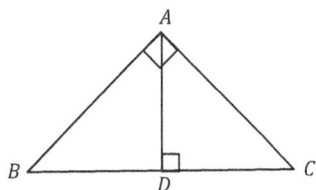

For any right triangle $\triangle ABC$ with $m \angle A = 90°$, the altitude (height) to the hypotenuse $\overline{BC}$ separates the triangle $\triangle ABC$ into two triangles $\triangle ABD$ and $\triangle ACD$, where $\triangle ABC \sim \triangle DBA \sim \triangle DAC$.

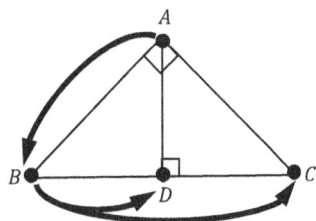

(1) Since $\triangle ABC \sim \triangle DBA$, $\dfrac{AB}{DB} = \dfrac{BC}{BA}$.

Therefore, $\boxed{AB^2 = BD \cdot BC}$

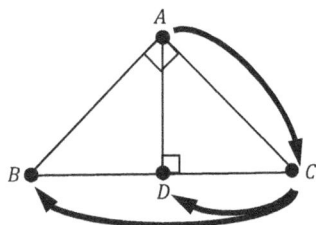

(2) Since $\triangle ABC \sim \triangle DAC$, $\dfrac{AC}{CD} = \dfrac{BC}{AC}$.

Therefore, $\boxed{AC^2 = CD \cdot CB}$

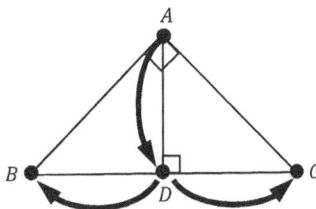

(3) Since $\triangle DBA \sim \triangle DAC$, $\dfrac{AD}{CD} = \dfrac{BD}{AD}$.

Therefore, $\boxed{AD^2 = DB \cdot DC}$

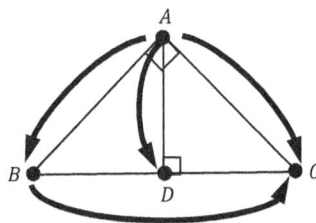

(4) Since the area of $\triangle ABC$ is $\frac{1}{2} \cdot AB \cdot AC = \frac{1}{2} \cdot BC \cdot AD$,

$AB \cdot AC = AD \cdot BC$

Or, since $\triangle ABC \sim \triangle DAC$, $\dfrac{BC}{AC} = \dfrac{BA}{AD}$.

Therefore, $\boxed{AB \cdot AC = AD \cdot BC}$

8-3 Parallel Lines

1. Triangles and Parallels

(1) The Proportionality Theorem and its Converse

1) The Basic Proportionality Theorem

For a triangle $\triangle ABC$, let D and E be points on $\overrightarrow{AB}$ and $\overrightarrow{AC}$.

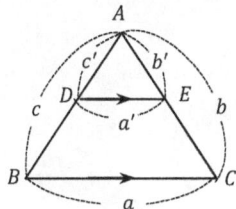

① If $\overline{DE} /\!/ \overline{BC}$, then $\dfrac{a'}{a} = \dfrac{b'}{b} = \dfrac{c'}{c}$.

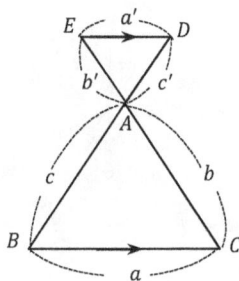

② If $\overline{DE} /\!/ \overline{BC}$, then $\dfrac{a'}{a} = \dfrac{b'}{b} = \dfrac{c'}{c}$.

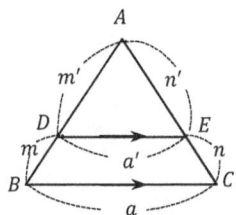

③ If $\overline{DE} /\!/ \overline{BC}$, then $\dfrac{m'}{m} = \dfrac{n'}{n} \neq \left(\dfrac{a'}{a}\right)$.

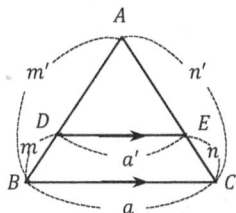

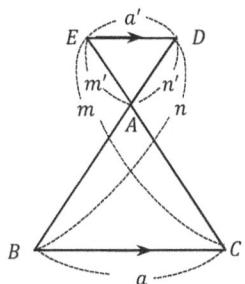

④ If $\overline{DE} /\!/ \overline{BC}$, then $\dfrac{m'}{m} = \dfrac{n'}{n} \neq \left(\dfrac{a'}{a}\right)$.

⑤ If $\overline{DE} /\!/ \overline{BC}$, then $\dfrac{m'}{m} = \dfrac{n'}{n} \neq \left(\dfrac{a'}{a}\right)$.

2) **The Converse of the Basic Proportionality theorem**

For a triangle $\triangle ABC$, let D and E be points on $\overrightarrow{AB}$ and $\overrightarrow{AC}$.

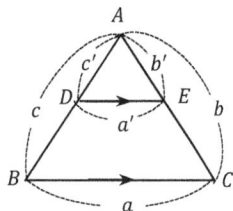

① If $\dfrac{a'}{a} = \dfrac{b'}{b} = \dfrac{c'}{c}$, then $\overline{DE} \mathbin{/\!/} \overline{BC}$.

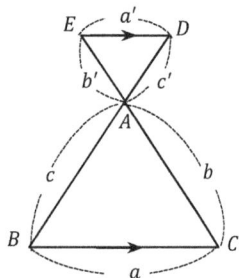

② If $\dfrac{a'}{a} = \dfrac{b'}{b} = \dfrac{c'}{c}$, then $\overline{DE} \mathbin{/\!/} \overline{BC}$.

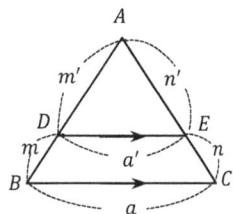

③ If $\dfrac{m'}{m} = \dfrac{n'}{n}$, then $\overline{DE} \mathbin{/\!/} \overline{BC}$.

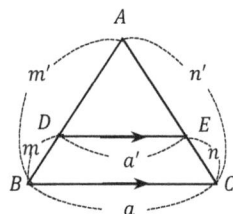

④ If $\dfrac{m'}{m} = \dfrac{n'}{n}$, then $\overline{DE} \mathbin{/\!/} \overline{BC}$.

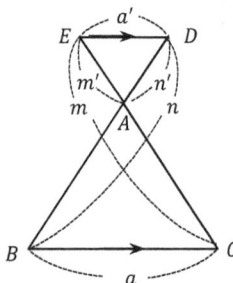

⑤ If $\dfrac{m'}{m} = \dfrac{n'}{n}$, then $\overline{DE} \mathbin{/\!/} \overline{BC}$.

(2) The Angle Bisector Theorems

    1) Bisector of an Interior Angle

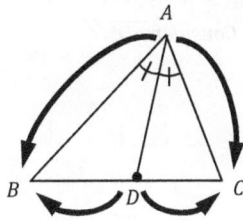

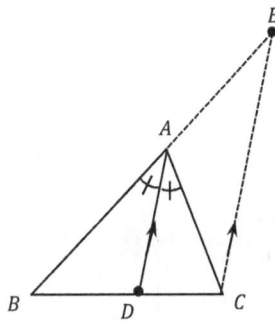

For a triangle $\triangle ABC$,

if $\overline{AD}$ is the bisector of the interior angle at A, and D is

on $\overline{BC}$, then $\dfrac{AB}{AC} = \dfrac{DB}{DC}$.

($\because$ Since $\overline{AD}$ is the bisector of $\angle A$, $\angle BAD \cong \angle DAC$.

Consider $\overline{DA} \mathbin{/\!/} \overline{CE}$.

Then $\angle DAC \cong \angle ECA$ by alternate interior angles.

Since $\angle BAC$ is an exterior angle of $\angle EAC$,

$m\angle BAC = m\angle ACE + m\angle AEC$.

Since $m\angle BAC = m\angle BAD + m\angle DAC$ and

$m\angle BAD = m\angle DAC$, $m\angle ACE = m\angle AEC$.

So, $\triangle ACE$ is an isosceles triangle. $AC = AE$

Since $\overline{DA} \mathbin{/\!/} \overline{CE}$, $\dfrac{BD}{DC} = \dfrac{BA}{AE}$

Since $AC = AE$, $\dfrac{BD}{DC} = \dfrac{BA}{AC}$.

Therefore, $\dfrac{AB}{AC} = \dfrac{DB}{DC}$.)

    2) Bisector of an Exterior Angle

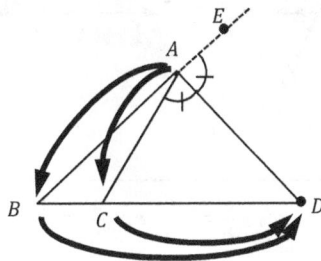

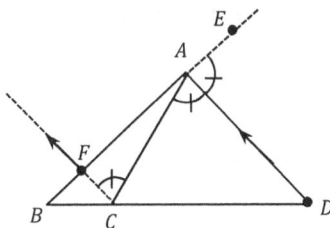

For a triangle $\triangle ABC$,

if $\overline{AD}$ is the bisector of the exterior angle at, A and D is on $\overrightarrow{BC}$,

then $\dfrac{AB}{AC} = \dfrac{BD}{CD}$.

($\because$ Since $\overline{AD}$ is the bisector of $\angle EAC$, $\angle EAD \cong \angle DAC$.

Consider $\overline{CF} \mathbin{/\!/} \overline{DA}$.

Then $\angle DAC \cong \angle FCA$ by alternate interior angles

and $\angle EAD \cong \angle ACF$

Since $m\angle EAC = m\angle AFC + m\angle ACF$, $\angle AFC \cong \angle ACF$.

So, $\triangle AFC$ is an isosceles triangle. $AF = AC$

Since $\overline{CF} \mathbin{/\!/} \overline{DA}$, $\dfrac{BA}{FA} = \dfrac{BD}{CD}$.

Since $AF = AC$, $\dfrac{BA}{AC} = \dfrac{BD}{CD}$.

Therefore, $\dfrac{AB}{AC} = \dfrac{BD}{CD}$.)

Note :

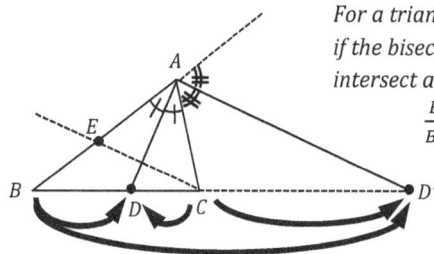

For a triangle $\triangle ABC$,
if the bisectors $\overline{AD}$ *and* $\overline{AD'}$ *of the interior and exterior angles at A intersect at points D and D', respectively, then*
$$\frac{BD}{BD'} = \frac{CD}{CD'}. \quad (\because Consider\ \overline{AD'}\ /\!/\ \overline{EC}\ )$$

2. Parallels and Transversals

(1) The Segment Proportionality Theorem

> Two or more parallel lines divide two transversals proportionally.
>
> The intercepted segments on the two transversals are in proportion.

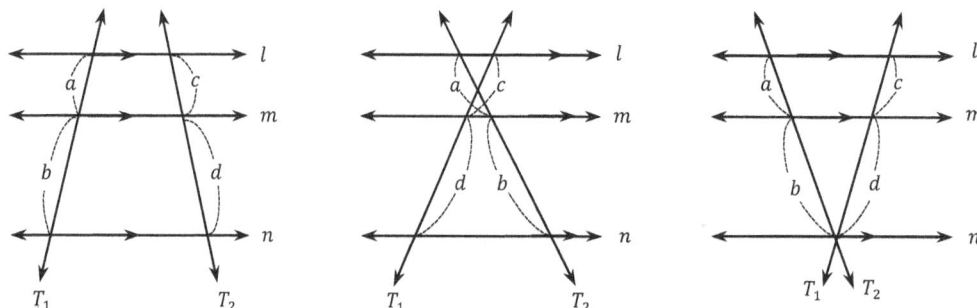

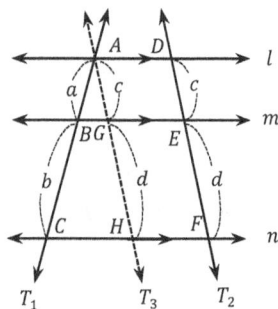

> If $l\,/\!/\,m\,/\!/\,n$, then $\dfrac{a}{b} = \dfrac{c}{d}$ or $\dfrac{a}{c} = \dfrac{b}{d}$.

$\because$ Let two transversals T_1 and T_2 cut parallels l, m, and n in A, B, C, D, E, F, respectively.

Consider a line T_3 passing through a point A and running parallel to T_2.

Let T_3 cut m and n in G and H, respectively.

Then $\overline{AG} \cong \overline{DE}$ and $\overline{GH} \cong \overline{EF}$.

In a triangle $\triangle ACH$, $\dfrac{a}{b} = \dfrac{c}{d}$ ($\because\ \overline{BG}\,/\!/\,\overline{CH}$).

(2) For a trapezoid $\square ABCD$ with $\overline{AD} /\!/ \overline{BC}$,

if $\overline{EF} /\!/ \overline{BC}$ and $AD = a$, $BC = b$, $AE = m$, $EB = n$, then $EF = \frac{an+bm}{m+n}$.

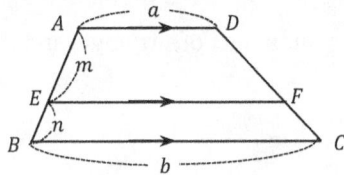

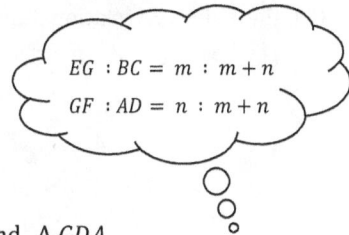

$EG : BC = m : m + n$

$GF : AD = n : m + n$

Imagine triangles $\triangle ABC$ and $\triangle CDA$.

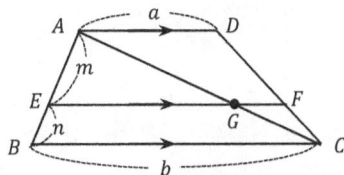

In $\triangle ABC$, $\frac{m}{EG} = \frac{m+n}{b}$ $\quad \therefore$ $(m + n)EG = bm$; $EG = \frac{bm}{m+n}$

In $\triangle CDA$, $\frac{n}{GF} = \frac{m+n}{a}$ $\quad \therefore$ $(m + n)GF = an$; $GF = \frac{an}{m+n}$

Since $EF = EG + GF$, $EF = \frac{bm}{m+n} + \frac{an}{m+n} = \frac{an+bm}{m+n}$.

$\frac{a}{b} \neq \frac{c}{d}$ ($\because \square ABCD$ and $\square AEFD$ are not similar.)

If the corresponding angles are congruent

and the corresponding sides are in proportion,

then the correspondence is a similarity.

(3) $\overline{AB} /\!/ \overline{CD} /\!/ \overline{EF}$ $\quad \Rightarrow \quad CD = \frac{ab}{a+b}$ and $\frac{BD}{DF} = \frac{a}{b}$

($\because$ Since $\triangle ABC \sim \triangle FEC$, $\frac{BC}{CE} = \frac{a}{b}$

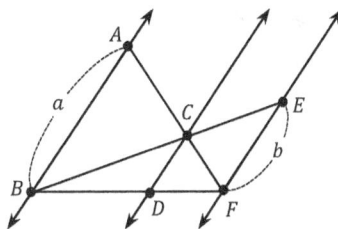

In $\triangle BEF$, $\frac{BC}{BE} = \frac{BD}{BF} = \frac{a}{a+b}$ In $\triangle FAB$, $\frac{FC}{FA} = \frac{FD}{FB} = \frac{b}{a+b}$

Since $\triangle BCD \sim \triangle BEF$, $\frac{CD}{EF} = \frac{a}{a+b}$; $CD = \frac{a \cdot EF}{a+b} = \frac{ab}{a+b}$

Since $\triangle FCD \sim \triangle FAB$, $\frac{CD}{AB} = \frac{b}{a+b}$; $CD = \frac{b \cdot AB}{a+b} = \frac{ab}{a+b}$

Therefore, $CD = \frac{ab}{a+b}$

In $\triangle ABC$ and $\triangle FEC$, $\frac{BC}{CE} = \frac{a}{b}$ In $\triangle BEF$, $\frac{BC}{CE} = \frac{BD}{DF}$

Therefore, $\frac{BD}{DF} = \frac{a}{b}$

8-4 Midpoints

1. Similar Triangles

(1) Consider the relationship of the segment connecting two midpoints of a triangle and the opposite side.

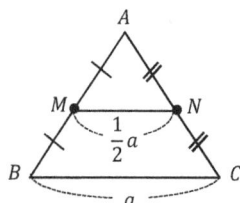

For triangle $\triangle ABC$,

let M and N be the midpoints of $\overline{AB}$ and $\overline{AC}$, respectively.

Then, $\overline{MN} /\!/ \overline{BC}$ and $MN = \frac{1}{2}BC$.

($\because$ Consider $\triangle ABC$ and $\triangle AMN$.

Since $\dfrac{AM}{AB} = \dfrac{AN}{AC} = \dfrac{1}{2}$, the two pairs of corresponding sides are proportional.

Since $\angle A$ is a common angle for both triangles, the included angles are congruent.

So, $\triangle ABC \sim \triangle AMN$.

Thus, $\angle AMN \cong \angle ABC$.

Since the corresponding angles are congruent, $\overline{MN} /\!/ \overline{BC}$.

Since $\triangle ABC \sim \triangle AMN$, $\dfrac{MN}{BC} = \dfrac{AM}{AB} = \dfrac{1}{2}$.

So, $2MN = BC$.

Therefore, $MN = \frac{1}{2}BC$.)

Note :

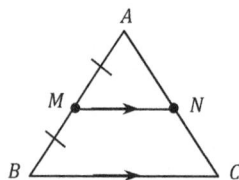

If M is the midpoint of $\overline{AB}$ *and* $\overline{MN} /\!/ \overline{BC}$,

Then N is the midpoint of $\overline{AC}$.

That is, $AM = MB$ and $\overline{MN} /\!/ \overline{BC} \Rightarrow AN = NC$.

($\because$ Since $\overline{MN} /\!/ \overline{BC}$, $\dfrac{AM}{MB} = \dfrac{AN}{NC}$

Since $\dfrac{AM}{MB} = \dfrac{1}{1}$, $AN = NC$.)

(2) Consider a trapezoid □ $ABCD$ with $\overline{AD} \parallel \overline{BC}$.

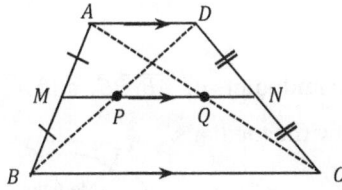

If M and N are the midpoints of $\overline{AB}$ and $\overline{DC}$, respectively,

then ① $\overline{AD} \parallel \overline{MN} \parallel \overline{BC}$,

② $MN = \frac{1}{2}(AD + BC)$, and

③ $PQ = \frac{1}{2}(BC - AD)$, where $BC > AD$.

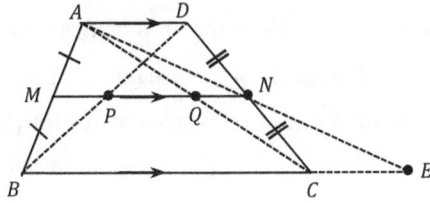

$MN = MP + PN$

$= \frac{1}{2}AD + \frac{1}{2}BC$

$= \frac{1}{2}(AD + BC)$

(∵ In $\triangle ADN$ and $\triangle ECN$, $\angle AND \cong \angle CNE$ (vertical angles).

Since $\overline{AD} \parallel \overline{CE}$, $\angle ADN \cong \angle ECN$ (alternate interior angles).

Since N is the midpoint of $\overline{DC}$, $DN = NC$.

So, $\triangle ADN \cong \triangle ECN$ by ASA Postulate.

∴ $AN = NE$, $AD = CE$

In $\triangle ABE$, $AM = MB$ and $AN = NE$.

So, $\overline{MN} \parallel \overline{BE}$ and $MN = \frac{1}{2}BE = \frac{1}{2}(BC + CE) = \frac{1}{2}(BC + AD)$ by (1)

Therefore, $\overline{AD} \parallel \overline{MN} \parallel \overline{BC}$ and $MN = \frac{1}{2}(AD + BC)$.

Also, $PQ = MQ - MP = \frac{1}{2}BC - \frac{1}{2}AD = \frac{1}{2}(BC - AD)$.)

8-5 Centroid

1. Similar Triangles

(1) For a triangle $\triangle ABC$, let points $M_1, M_2,$ and M_3 be the midpoints of $\overline{AB}, \overline{BC},$ and $\overline{AC}$, respectively. Then the triangle $\triangle ABC$ is separated into six triangles by the medians.

Medians:
$\overline{AM_2}, \overline{BM_3},$ and $\overline{CM_1}$

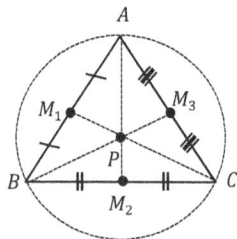

In this case,

the area of $\triangle ABM_2$ = the area of $\triangle ACM_2$,

the area of $\triangle ABM_3$ = the area of $\triangle BCM_3$, and

the area of $\triangle CAM_1$ = the area of $\triangle CBM_1$.

The intersection point P of the three medians is called the *centroid* or *center of gravity* of the triangle.

The distance between the centroid to each vertex is $\frac{2}{3}$ the length of each median.

That is, $AP : PM_2 = BP : PM_3 = CP : PM_1 = 2 : 1$

The area of $\triangle ABM_2 = \frac{1}{2} \cdot BM_2 \cdot h$

The area of $\triangle ACM_2 = \frac{1}{2} \cdot M_2C \cdot h$

Since $BM_2 = M_2C$,

the area of $\triangle ABM_2$ = the area of $\triangle ACM_2$.

Note

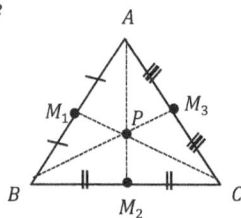

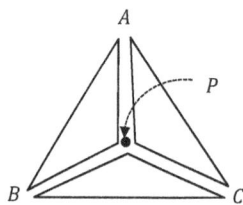

The area of $\triangle ABP$

$=$ The area of $\triangle BCP$

$=$ The area of $\triangle ACP$

$= \frac{1}{3} \times$ The area of $\triangle ABC$

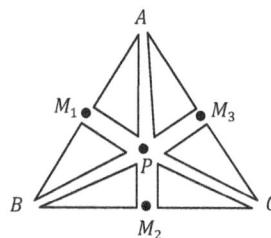

The area of $\triangle AM_1P$

$=$ The area of $\triangle BM_1P$

$=$ The area of $\triangle BM_2P$

$=$ The area of $\triangle CM_2P$

$=$ The area of $\triangle CM_3P$

$=$ The area of $\triangle AM_3P$

$= \frac{1}{6} \times$ The area of $\triangle ABC$

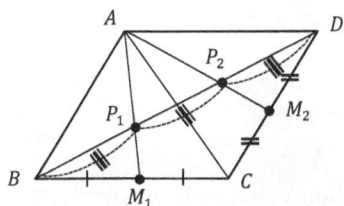

(2) For a parallelogram $\square ABCD$,

  let P_1 and P_2 be the centroids of $\triangle ABC$ of $\triangle ACD$,

  respectively. Then $BP_1 = P_1P_2 = P_2D$.

($\because$ Since $\square ABCD$ is a parallelogram,

the two diagonals of $\square ABCD$ bisect each other.

$\therefore$ $BQ = QD$

Since P_1 is the centroid of $\triangle ABC$, $BP_1 : P_1Q = 2 : 1$

Since P_2 is the centroid of $\triangle ACD$, $DP_2 : P_2Q = 2 : 1$

Therefore, $BP_1 : P_1P_2 : P_2D = 2 : 2 : 2$

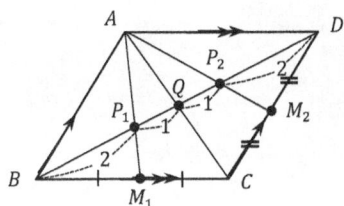

$\therefore$ $BP_1 = P_1P_2 = P_2D$)

8-6 Areas and Volumes

1. Perimeter and Area

For any two similar polygons,

if the ratio of the lengths of their corresponding sides is $m : n$,

then the ratio of their perimeters is $m : n$ and the ratio of their areas is $m^2 : n^2$.

($\because$ Suppose $\triangle ABC \sim \triangle DEF$.

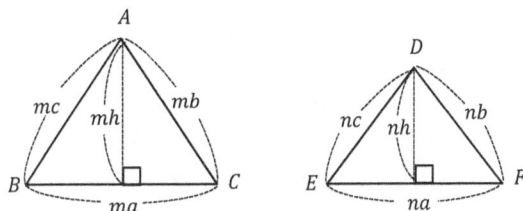

Since the perimeter P_1 of $\triangle ABC$ is $P_1 = ma + mb + mc = m(a + b + c)$ and

the perimeter P_2 of $\triangle DEF$ is $P_2 = na + nb + nc = n(a + b + c)$,

$P_1 : P_2 = m(a + b + c) : n(a + b + c) = m : n$

Since the area A_1 of $\triangle ABC$ is $A_1 = \frac{1}{2} \cdot ma \cdot mh = \frac{1}{2} \cdot m^2 ah$ and

the area A_2 of $\triangle DEF$ is $A_2 = \frac{1}{2} \cdot na \cdot nh = \frac{1}{2} \cdot n^2 ah$,

$A_1 : A_2 = \frac{1}{2} \cdot m^2 ah : \frac{1}{2} \cdot n^2 ah = m^2 : n^2$)

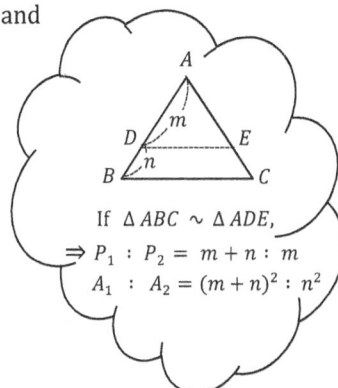

If $\triangle ABC \sim \triangle ADE$,
$\Rightarrow P_1 : P_2 = m + n : m$
$A_1 : A_2 = (m + n)^2 : n^2$

2. Surface Area and Volume

For any two similar solids,

if the ratio of the lengths of their corresponding lateral edges is $m : n$,

then the ratio of their surface areas is $m^2 : n^2$ and the ratio of their volumes is $m^3 : n^3$.

(∵ Suppose two rectangular prisms are similar.

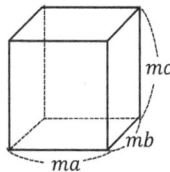
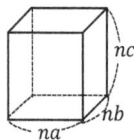

Figure 1   Figure 2

Then the surface area S_1 of Figure 1 is

$S_1 = 2 \cdot ma \cdot mb + 2 \cdot mb \cdot mc + 2 \cdot ma \cdot mc = 2\,m^2(ab + bc + ac)$ and

the surface area S_2 of Figure 2 is

$S_2 = 2 \cdot na \cdot nb + 2 \cdot nb \cdot nc + 2 \cdot na \cdot nc = 2\,n^2(ab + bc + ac)$.

Therefore, $S_1 : S_2 = 2\,m^2(ab + bc + ac) : 2\,n^2(ab + bc + ac) = m^2 : n^2$

The volume V_1 of Figure 1 is $V_1 = ma \cdot mb \cdot mc = m^3 abc$ and

the volume V_2 of Figure 2 is $V_2 = na \cdot nb \cdot nc = n^3 abc$.

Therefore, $V_1 : V_2 = m^3 abc : n^3 abc = m^3 : n^3$)

Exercises

#1 Given two similar polygons with lengths of sides or measures of angles as marked, find the specifies lengths/angles :

(1) $\triangle ABC \sim \triangle DEF$ Find a and b.

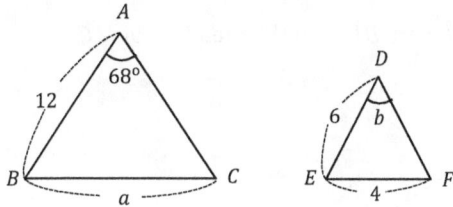

(2) $\triangle ABC \sim \triangle DEF$ Find $m \angle D$ and $m \angle F$.

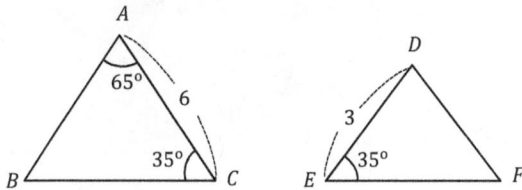

(3) $\square ABCD \sim \square EFGH$ Find AB and $m \angle A$.

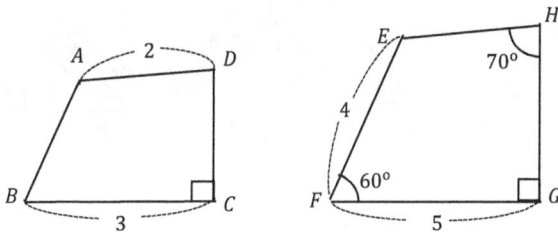

(4) $\triangle ABC \sim \triangle DEF$ Find a and b.

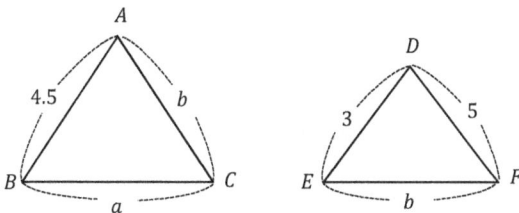

(5) ▱ $ABCD \sim$ ▱ $EFGH$ with $AB : EF = 2 : 3$ Find the perimeter of ▱ $EFGH$.

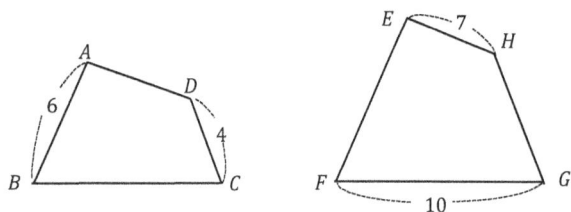

(6) $\triangle ABC \sim \triangle EFD$ Find the lengths of $\overline{AB}$ and $\overline{DF}$ and the measure of $\angle C$.

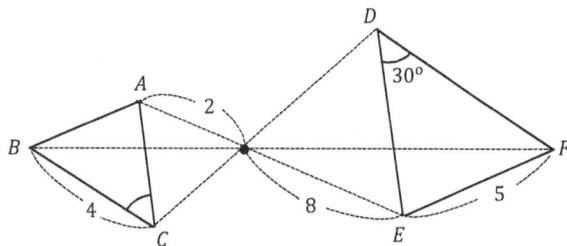

#2 Find the value of a for the following triangles :

(1)

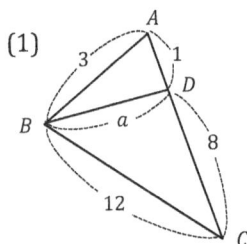

(2)

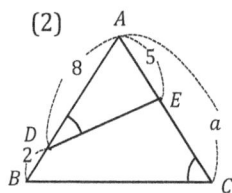

(3)

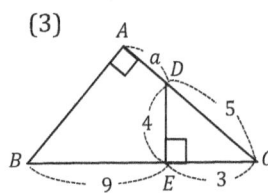

(4)

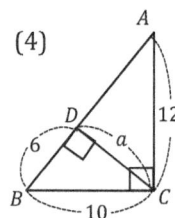

(5)

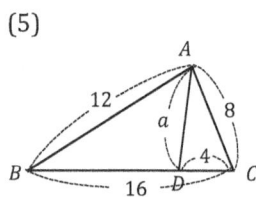

(6)

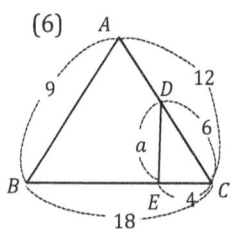

(7)

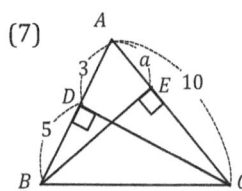

(8)

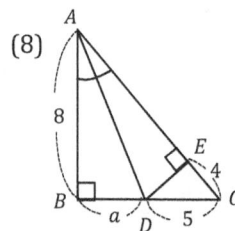

#3 $\overline{DE} /\!/ \overline{BC}$. Find the value of a and b for the following triangles :

(1)

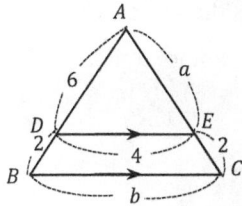

(2)

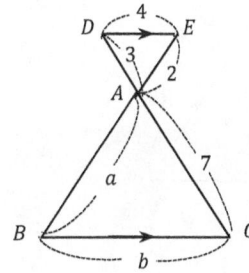

(3)

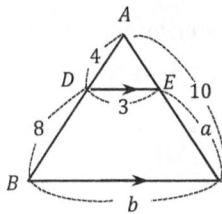

(4)

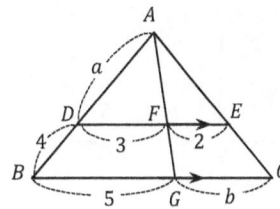

#4 For each triangle $\triangle ABC$, $\overline{AD}$ is the bisector of $\angle A$. Find the value of a.

(1)

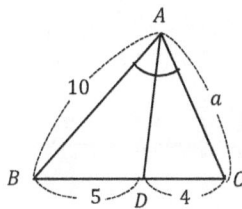

(2)

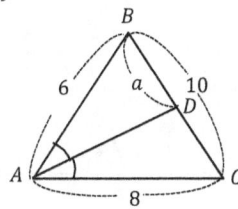

(3)

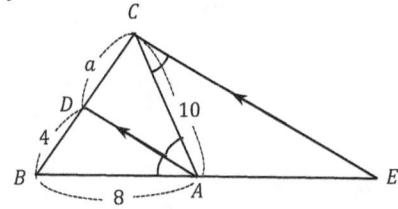

#5 For each triangle $\triangle ABC$, $\overline{AD}$ is the bisector of an exterior angle at A. Find the value of a.

(1)

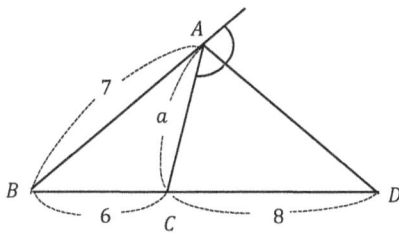

(2)

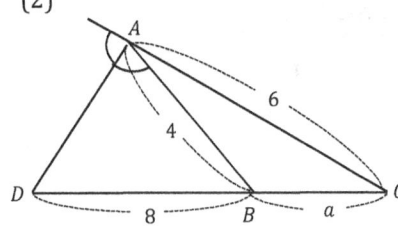

#6 Answer the following :

(1)

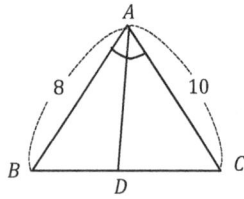

In triangle $\triangle ABC$, $\overline{AD}$ is the bisector of $\angle A$.

The area of $\triangle ABD$ is 12 square units.

Find the area of $\triangle ADC$.

(2)

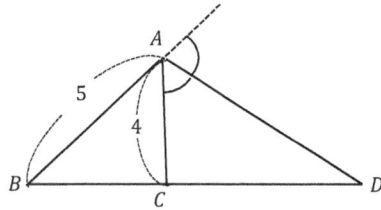

In triangle $\triangle ABC$,

$\overline{AD}$ is the bisector of an exterior angle at A.

Find the ratio of the areas of $\triangle ABD$ and $\triangle ACD$.

#7 $l /\!/ m /\!/ n$ Find the value of a.

(1)

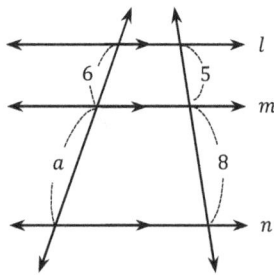

(2)

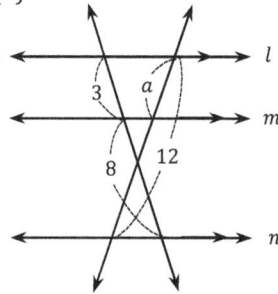

(3)

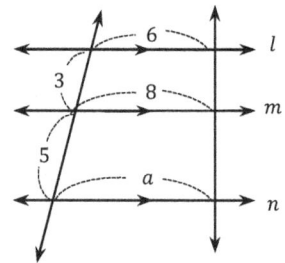

(4)

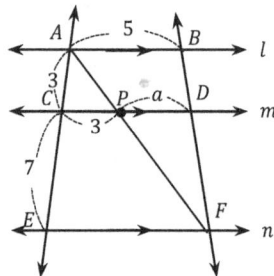

(5)

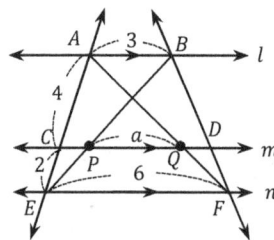

(6)

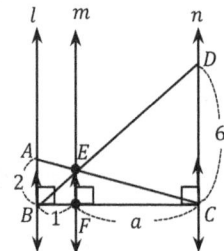

#8 For triangles $\triangle ABC$, answer the questions :

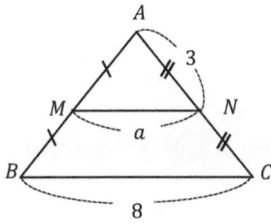

(1) M and N are midpoints of $\overline{AB}$ and $\overline{AC}$, respectively. Find the value of a and the length of $\overline{AC}$.

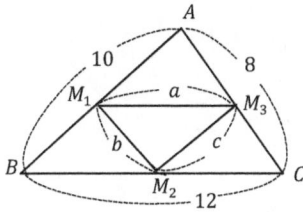

(2) M_1, M_2, and M_3 are midpoints of $\overline{AB}$, $\overline{BC}$, and $\overline{AC}$, respectively. Find the value of $a + b + c$.

#9 For parallelograms $\square\ ABCD$ with $\overline{AD}/\!/ \overline{BC}$, M and N are midpoints of $\overline{AB}$ and $\overline{DC}$. Find the value of a.

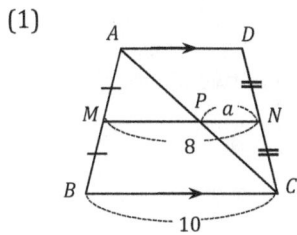

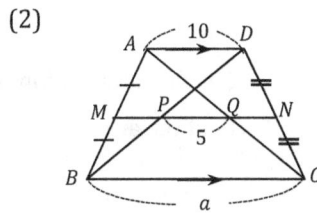

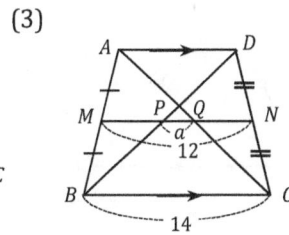

(1)

(2)

(3)

#10 The point P is the centroid of triangles $\triangle ABC$. Find the values of a and b .

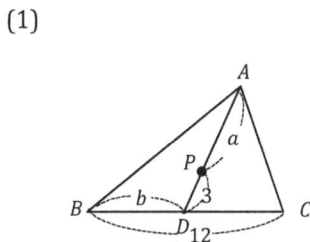

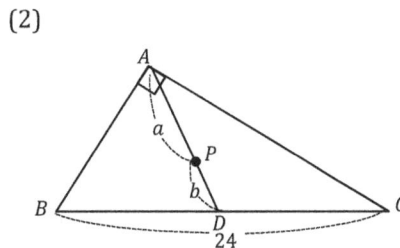

(1)

(2)

#11 Points P and Q are the centroids of the triangles $\triangle ABC$ and $\triangle PBC$. Answer the questions.

(1)

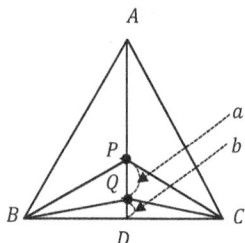

$AD = 24$.
Find the lengths of a and b, where $a = PQ, b = QD$.

(2)

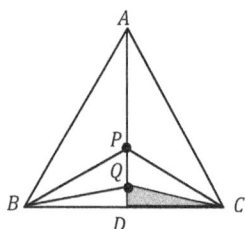

The area of $\triangle QDC$ is 5.
Find the area of $\triangle ABC$.

#12 For parallelograms $\square\, ABCD$ with areas of 72 in.^{2}, answer the following :

(1)

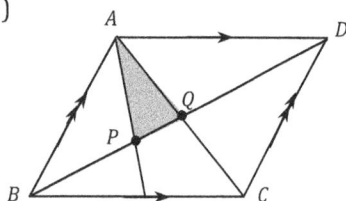

P is the centroid of $\triangle ABC$ and

Q is the intersection point of two diagonals of $\square\, ABCD$.

Find the area of $\triangle APQ$.

(2)

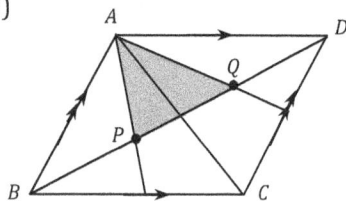

P is the centroid of $\triangle ABC$ and

Q is the centroid of $\triangle ACD$.

Find the area of $\triangle APQ$.

#13 For two similar polygons in each question, find the ratio of their areas.

(1)

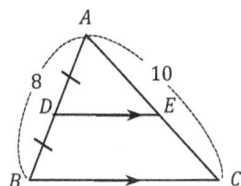

(2)

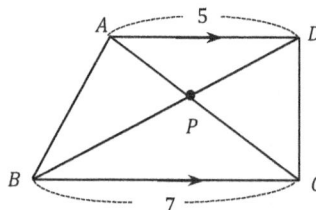

#14 Answer the following :

(1)

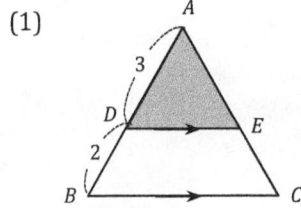

For triangle $\triangle ABC$,
$\overline{DE} \mathbin{/\!/} \overline{BC}$ and the area of $\triangle ADE$ is 30 in.^{2}
Find the area of $\square DBCE$.

(2)

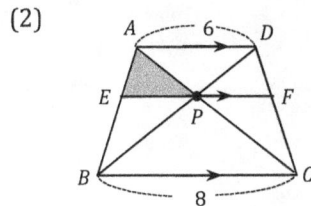

For trapezoid $\square ABCD$,
$\overline{AD} \mathbin{/\!/} \overline{EF} \mathbin{/\!/} \overline{BC}$.

The area of $\triangle AEP$ is 10 inch^2.

Find the area of $\triangle ABC$.

#15 For two similar cones A and B, $h_1 : h_2 = 3 : 5$. Answer the following :

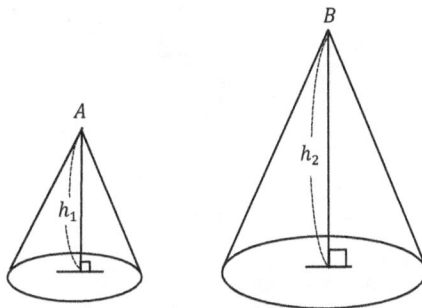

(1) The area S_1 of the lateral face of A is 144 in.^{2} Find the area S_2 of the lateral face of B.

(2) The volume V_1 of A is 216 in.^{3} Find the volume V_2 of B.

The Pythagorean Theorem

Chapter 9 The Pythagorean Theorem

CHAPTER
9

9-1 The Pythagorean Theorem

1. The Pythagorean Theorem
2. Proofs of the Pythagorean Theorem
    (1) Pythagorean Theorem Proof
    (2) Euclid's Proof of the Pythagorean Theorem
    (3) Bhaskara's Proof of the Pythagorean Theorem
3. The Converse of the Pythagorean Theorem
4. The Relationship of the Sides and Angles of a Triangle

9-2 Applications of the Pythagorean Theorem

1. Right Triangles
    (1) Similarity
    (2) Diagonals
    (3) Half Circles
    (4) Special Triangles : $45°$- $45°$- $90°$, $30°$- $60°$- $90°$
    (5) Distance
2. Polygons
    (1) Quadrilaterals
    (2) Rectangles
    (3) Regular Triangles
3. Solids
    (1) Right Rectangular Prisms
    (2) Regular Rectangular Prisms ; Cubes
    (3) Regular Rectangular Pyramids
    (4) Regular Triangular Pyramids
    (5) Cones
    (6) Spheres

Chapter 9. The Pythagorean Theorem

9-1   The Pythagorean Theorem

1. The Pythagorean Theorem

In a right triangle, the square of the length of the hypotenuse is equal to the sum of the squares of the lengths of the legs.

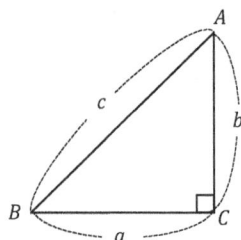

$$a^2 + b^2 = c^2$$

hypotenuse   legs

right angle

The hypotenuse is the longest side opposite the right angle.

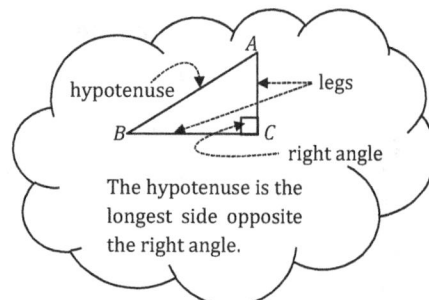

If the lengths of any two sides of a right triangle are given, we can find the length of the other side by the Pythagorean Theorem.

$$x^2 = a, \quad a \geq 0$$
$$\Rightarrow x = \pm\sqrt{a}$$

$a^2 + b^2 = c^2 \quad \Rightarrow \quad c = \pm\sqrt{a^2 + b^2}$

But c is a length of a triangle.

So $c > 0$.

Therefore, $c = +\sqrt{a^2 + b^2} = \sqrt{a^2 + b^2}$

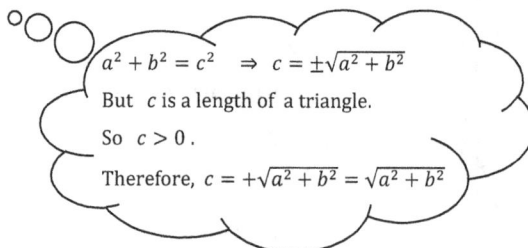

2. Proofs of the Pythagorean Theorem

(1) Pythagorean Theorem Proof

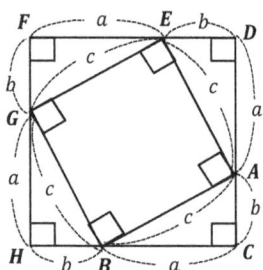

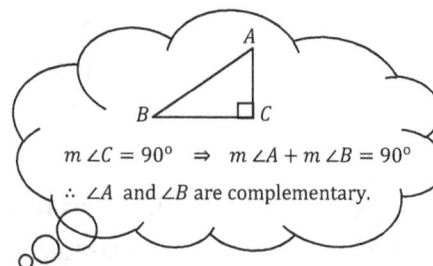

$m\angle C = 90^\circ \quad \Rightarrow \quad m\angle A + m\angle B = 90^\circ$

$\therefore \ \angle A$ and $\angle B$ are complementary.

From a right triangle $\triangle ABC$, imagine a square $\square CDFH$ with sides of length $a + b$

Then $\triangle ABC$, $\triangle AED$, $\triangle EGF$, and $\triangle GBH$ are right triangles with legs a and b.

By SAS Postulate, $\triangle ABC \cong \triangle AED \cong \triangle EGF \cong \triangle GBH$.

The quadrilateral $\square ABGE$ formed by the four hypotenuses is a square with sides of length c.

Since the acute angles of a right triangle are complementary, $m\angle HBG + m\angle CBA = 90^\circ$.

Since $m\angle HBG + m\angle GBA + m\angle CBA = 180^\circ$, $m\angle GBA = 90^\circ$.

Similarly, $m\angle BAE = m\angle AEG = m\angle EGB = 90^\circ$.

So $\square ABGE$ is a square.

Therefore, the area of the large square $\square CDFH$ is equal to the area of the small square $\square ABGE$ added to the sum of the areas of the four congruent triangles $\triangle ABC$, $\triangle AED$, $\triangle EGF$, and $\triangle GBH$.

This gives $(a+b)^2 = c^2 + 4 \cdot (\frac{1}{2}ab)$; $a^2 + 2ab + b^2 = c^2 + 2ab$.

Therefore, $a^2 + b^2 = c^2$.

(2) Euclid's Proof of the Pythagorean Theorem

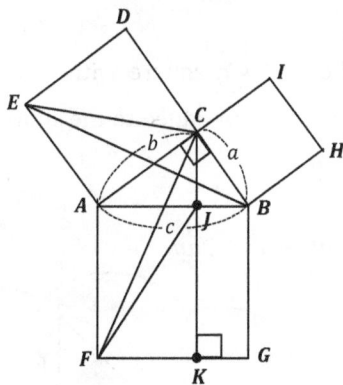

From a right triangle $\triangle ABC$ with $m \angle C = 90°$, imagine three squares with side lengths of $a, b,$ and c.

Then, the area of the square with length c (the length of hypotenuse of the triangle) is equal to the sum of the areas of the other two squares. ; $c^2 = a^2 + b^2$

(∵ In $\triangle EAB$ and $\triangle CAF$, $EA = AC$, $AB = AF$, $\angle EAB \cong \angle CAF$

So, $\triangle EAB \cong \triangle CAF$ by SAS Postulate.

Since $\overline{DB} \parallel \overline{EA}$, the area of $\triangle EAC$ = the area of $\triangle EAB$ (∵ congruent heights and bases).

Since $\overline{AF} \parallel \overline{CK}$, the area of $\triangle CAF$ = the area of $\triangle JAF$ (∵ congruent heights and bases).

Therefore, the area of $\triangle EAC$ = the area of $\triangle JAF$

Therefore, the area of $\square ACDE$ = the area of $\square AFKJ$.

Similarly, the area of $\square BHIC$ = the area of $\square JKGB$.

Hence, the area of $\square AFGB$ is equal to the sum of the area of $\square ACDE$ and the area of $\square CBHI$.

This means $c^2 = a^2 + b^2$.)

(3) Bhaskara's Proof of the Pythagorean Theorem

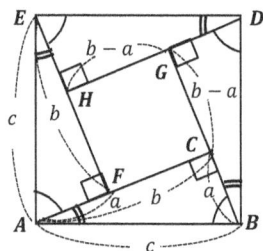

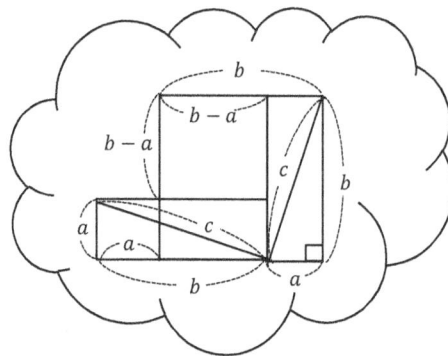

(∵ Since $\triangle ABC \cong \triangle BDG \cong \triangle DEH \cong \triangle EAF$,

$\square\ ABDE$ is a square with side lengths of c and $\square\ CGHF$ is a square with side lengths of $b - a$. Therefore, the area of $\square\ ABDE$ is equal to the area of $\square\ CGHF$ added to the sum of the areas of the four congruent triangles.

This means $c^2 = (b - a)^2 + 4\left(\frac{1}{2}ab\right)$; $c^2 = b^2 - 2ab + a^2 + 2ab$

Hence, $c^2 = a^2 + b^2$.)

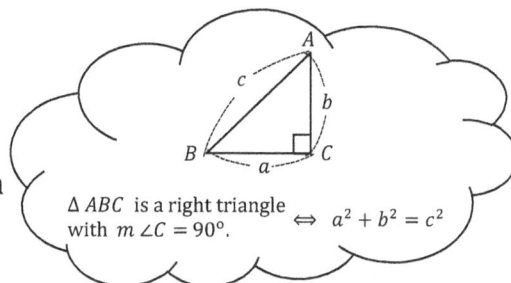

$\triangle ABC$ is a right triangle with $m\angle C = 90°.$ $\Leftrightarrow$ $a^2 + b^2 = c^2$

3. The Converse of the Pythagorean Theorem

If the square of one side of a triangle is equal to the sum of the squares of the other two sides, then the triangle is a right triangle with its right angle opposite the longest side.

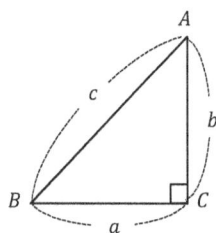

For a triangle $\triangle ABC$,

if $a^2 + b^2 = c^2$,

then $\triangle ABC$ is a right triangle with $m\angle C = 90°$.

(∵

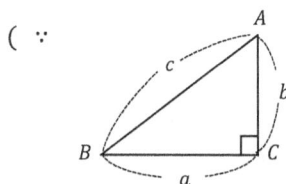

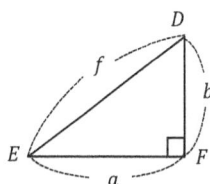

For a triangle $\triangle ABC$ with $a^2 + b^2 = c^2$,

imagine a right triangle $\triangle DEF$ with legs a and b, and hypotenuse f.

Since $\triangle DEF$ is a right triangle, $a^2 + b^2 = f^2$, by the Pythagorean Theorem.

Since $a^2 + b^2 = c^2$ is given, $c^2 = f^2$.

So, $c = \pm\sqrt{f^2} = \pm f$.

Since c is a length of a side, $c > 0$.

So, $c = f$.

Thus, $\triangle ABC \cong \triangle DEF$ by SSS Postulate.

So, $\angle C \cong \angle F$

Since $\angle F$ is a right angle, so is $\angle C$.

Therefore, the triangle $\triangle ABC$ is a right triangle with $m \angle C = 90°$.)

4. The Relationship of the Sides and Angles of a Triangle

For a triangle $\triangle ABC$ with the longest side of length c,

(1) $a^2 + b^2 > c^2 \Rightarrow m \angle C < 90°$; acute
(2) $a^2 + b^2 = c^2 \Rightarrow m \angle C = 90°$; right
(3) $a^2 + b^2 < c^2 \Rightarrow m \angle C > 90°$; obtuse

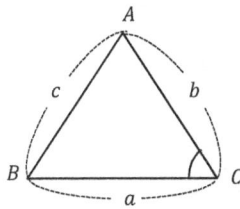 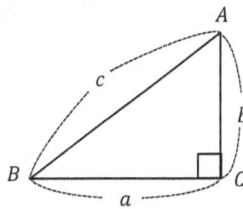 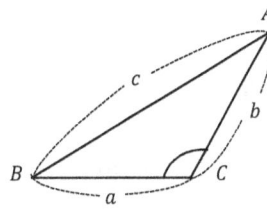

$\angle C$: acute $\angle C$: right $\angle C$: obtuse

9-2   Applications of the Pythagorean Theorem

1. Right Triangles

(1) Similarity

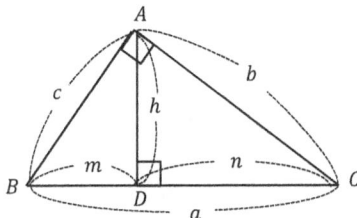

For a triangle $\triangle ABC$ with $m \angle A = 90°$, let $\overline{AD} \perp \overline{BC}$.

Then,   ① $a^2 = b^2 + c^2$, by the Pythagorean Theorem.

② $b^2 = na$, $c^2 = ma$, $h^2 = mn$, by similarity.

③ $ah = bc$, by congruent areas.

(2) Diagonals

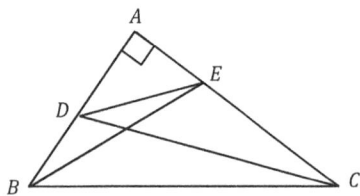

For a triangle $\triangle ABC$ with $m \angle A = 90°$,

$$BC^2 + DE^2 = BE^2 + CD^2$$

$$(\because\ BC^2 + DE^2 = (AB^2 + AC^2) + (AD^2 + AE^2) = (AB^2 + AE^2) + (AC^2 + AD^2) = BE^2 + CD^2\ )$$

(3) Half Circles

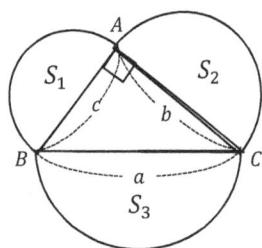

For a triangle $\triangle ABC$ with $m \angle A = 90°$,

Let S_1, S_2, and S_3 be the areas of the half circles with diameters c, b, and a, respectively.

Then,

① $S_1 + S_2 = S_3$ and

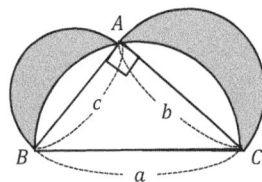

② the area of the shaded regions is equal to the area of $\triangle ABC$.

$(\because\ $ Since the triangle $\triangle ABC$ is a right triangle, $a^2 = b^2 + c^2$.

Since $S_1 = \frac{1}{2}\pi \left(\frac{c}{2}\right)^2$, $S_2 = \frac{1}{2}\pi \left(\frac{b}{2}\right)^2$, and $S_3 = \frac{1}{2}\pi \left(\frac{a}{2}\right)^2$, $S_1 + S_2 = \frac{1}{8}\pi c^2 + \frac{1}{8}\pi b^2$

Therefore, $S_1 + S_2 = \frac{1}{8}\pi (b^2 + c^2) = \frac{1}{8}\pi a^2 = S_3$.

The area of the shaded regions is

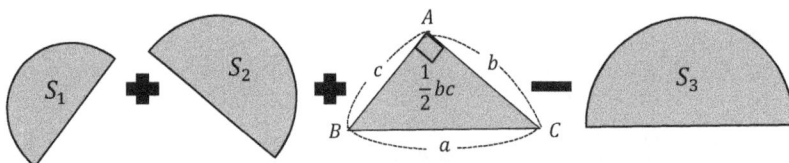

That is, the area of the shaded regions is

$S_1 + S_2 + \frac{1}{2}bc - S_3 = S_3 + \frac{1}{2}bc - S_3 = \frac{1}{2}bc = $ the area of $\triangle ABC$. $)$

(4) Special Triangles : 45°- 45°- 90°, 30°- 60°- 90°

1) **45°- 45°- 90° Triangles** $\Rightarrow$ $BC : CA : AB = 1 : 1 : \sqrt{2}$

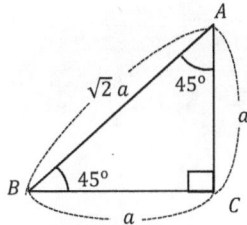

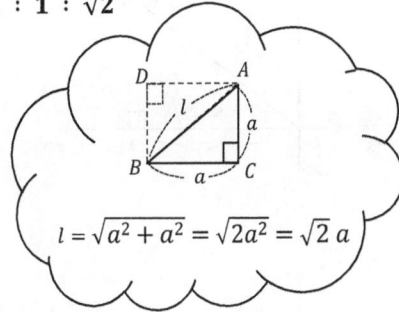

$$l = \sqrt{a^2 + a^2} = \sqrt{2a^2} = \sqrt{2}\,a$$

($\because$ Since $\angle A \cong \angle B$, $\triangle ABC$ is isosceles. $\therefore BC = CA$.

So, $AB^2 = a^2 + a^2 = 2a^2$. Therefore, $AB = \sqrt{2}\,a$.

Hence, $BC : CA : AB = a : a : \sqrt{2}\,a = 1 : 1 : \sqrt{2}$.)

2) **30°- 60°- 90° Triangles** $\Rightarrow$ $BC : CA : AB = 1 : \sqrt{3} : 2$

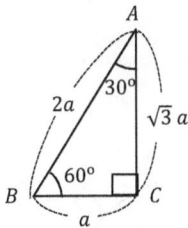

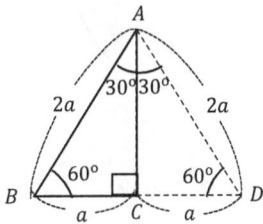

($\because$ Consider a congruent triangle $\triangle ADC$. $\triangle ABC \cong \triangle ADC$

Since $\triangle ABD$ is a regular triangle, $AB = 2a$.

$\therefore$ $AB^2 = BC^2 + AC^2$; $4a^2 = a^2 + AC^2$; $AC^2 = 3a^2$

So, $AC = \sqrt{3}a$

Therefore, $BC : CA : AB = a : \sqrt{3}a : 2a = 1 : \sqrt{3} : 2$)

(5) Distance

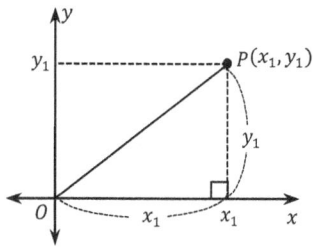

The distance between the origin O and a point $P(x_1, y_1)$

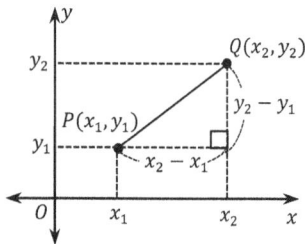

is $OP = \sqrt{x_1{}^2 + y_1{}^2}$

The distance between two points $P(x_1, y_1)$ and $Q(x_2, y_2)$

is $PQ = \sqrt{(x_2 - x_1)^2 + (y_2 - y_1)^2}$.

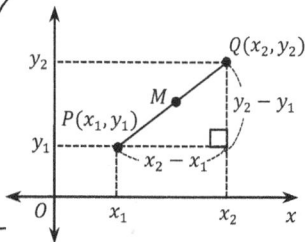

$$PQ = \sqrt{(x_2 - x_1)^2 + (y_2 - y_1)^2}$$
$$= \sqrt{(x_1 - x_2)^2 + (y_1 - y_2)^2}$$

The midpoint M between two points $P(x_1, y_1)$ and $Q(x_2, y_2)$ is

$$M = \left(\frac{x_1 + x_2}{2}, \frac{y_1 + y_2}{2} \right).$$

2. Polygons

(1) Quadrilaterals

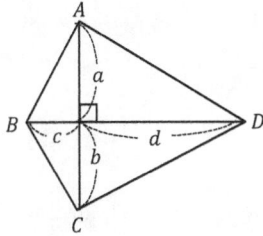

For a quadrilateral $\square ABCD$,

let the diagonals be perpendicular to each other.

Then, $AB^2 + CD^2 = AD^2 + BC^2$

($\because$ By the Pythagorean Theorem,

$a^2 + c^2 = AB^2$, $a^2 + d^2 = AD^2$, $b^2 + c^2 = BC^2$, and $b^2 + d^2 = CD^2$.

So, $AB^2 + CD^2 = a^2 + c^2 + b^2 + d^2$

and $AD^2 + BC^2 = a^2 + d^2 + b^2 + c^2$.

Therefore, $AB^2 + CD^2 = AD^2 + BC^2$.)

(2) Rectangles

1) Let P be a point in the interior of a rectangle $\square ABCD$.

Then, $AP^2 + CP^2 = BP^2 + DP^2$

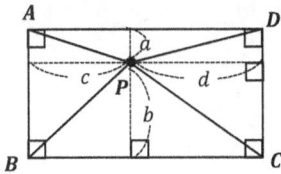

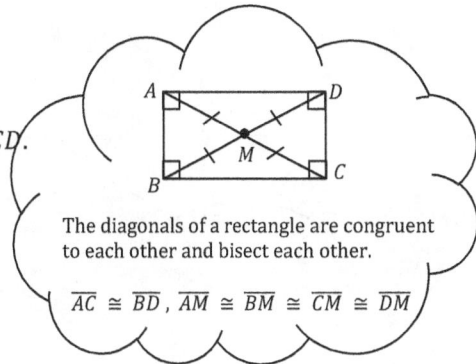

The diagonals of a rectangle are congruent to each other and bisect each other.

$\overline{AC} \cong \overline{BD}$, $\overline{AM} \cong \overline{BM} \cong \overline{CM} \cong \overline{DM}$

$(\because AP^2 + CP^2 = (a^2 + c^2) + (b^2 + d^2) = (b^2 + c^2) + (a^2 + d^2) = BP^2 + DP^2$)

2) Let l be the length of a diagonal of a rectangle $\square ABCD$ with two sides of lengths a and b.

Then, $l = \sqrt{a^2 + b^2}$.

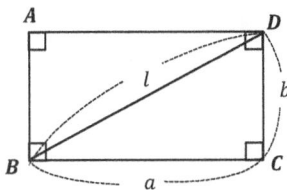

($\because \triangle DBC$ is a right triangle, $a^2 + b^2 = l^2$. Therefore, $l = \sqrt{a^2 + b^2}$.)

(3) Regular Triangles

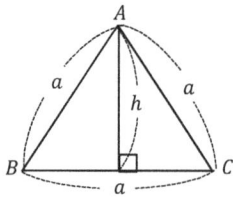

For a regular triangle $\triangle ABC$ with sides of lengths a, let h be the altitude (height) and S be the area of $\triangle ABC$. Then,

$$h = \frac{\sqrt{3}}{2}a \quad \text{and} \quad S = \frac{\sqrt{3}}{4}a^2.$$

(∵

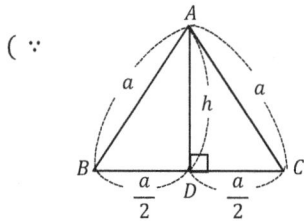

Since $\triangle ABD$ is a right triangle, $a^2 = h^2 + (\frac{a}{2})^2$.

So, $h^2 = a^2 - \left(\frac{a}{2}\right)^2 = \frac{3}{4}a^2$.

Therefore, $h = \frac{\sqrt{3}}{2}a$.

Since $S = \frac{1}{2}ah$ and $h = \frac{\sqrt{3}}{2}a$, $S = \frac{1}{2}a \cdot \frac{\sqrt{3}}{2}a = \frac{\sqrt{3}}{4}a^2$.)

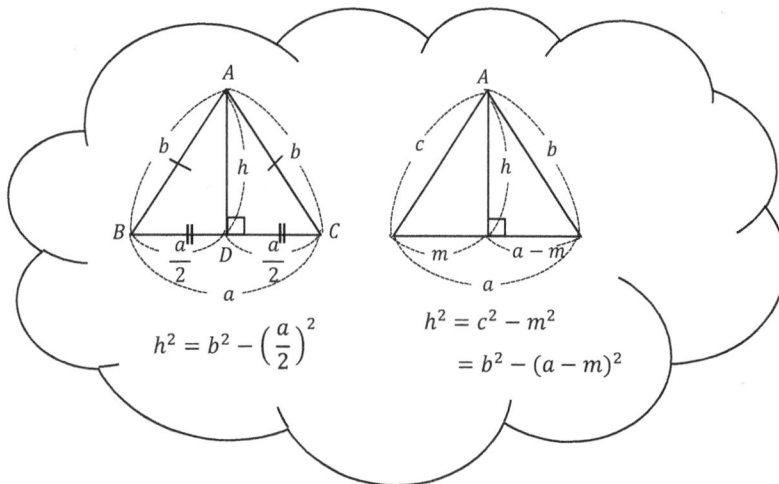

$$h^2 = b^2 - \left(\frac{a}{2}\right)^2$$

$$h^2 = c^2 - m^2 = b^2 - (a-m)^2$$

3. Solids

(1) Right Rectangular Prisms

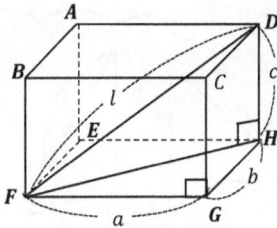

For a right rectangular prism with length a, width b, and height c, the length of a diagonal is

$$l = \sqrt{a^2 + b^2 + c^2}.$$

($\because$ Since $\square EFGH$ is a rectangle, $\triangle HFG$ is a right triangle.

$\therefore\ FH = \sqrt{a^2 + b^2}$

Since $\triangle DFH$ is a right triangle, $l^2 = FH^2 + c^2 = (a^2 + b^2) + c^2$.

Therefore, $l = \sqrt{a^2 + b^2 + c^2}$.)

(2) Regular Rectangular Prisms ; Cubes

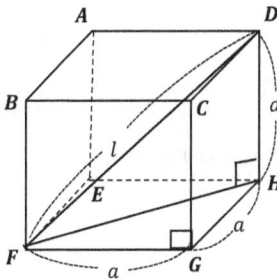

For a regular rectangular prism with sides of lengths a, the length of a diagonal is

$$l = \sqrt{a^2 + a^2 + a^2} = \sqrt{3a^2} = \sqrt{3}a.$$

(3) Regular Rectangular Pyramids

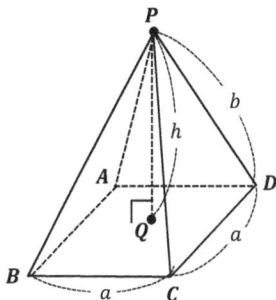

If a rectangular pyramid has a length b of its lateral edges and a length a of sides of its base, then the altitude (height) is

$$h = \sqrt{b^2 - \frac{a^2}{2}}.$$

($\because$ Since $\triangle BCD$ is a right triangle with $m\angle C = 90°$ and length a of legs, $BD = \sqrt{2}\,a$.

Since the diagonals of parallelograms bisect each other, $BQ = QD = \dfrac{\sqrt{2}\,a}{2}$.

Since $\triangle PQD$ is a right triangle, $b^2 = PQ^2 + (\dfrac{\sqrt{2}\,a}{2})^2$; $PQ^2 = b^2 - \dfrac{2a^2}{4} = b^2 - \dfrac{a^2}{2}$.

Therefore, the altitude is $h = \sqrt{b^2 - \dfrac{a^2}{2}}$.)

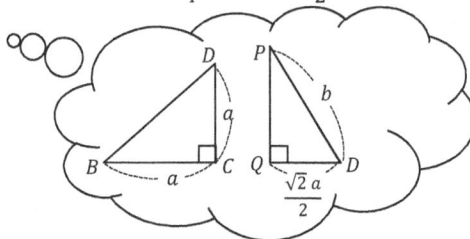

(4) Regular Triangular Pyramids

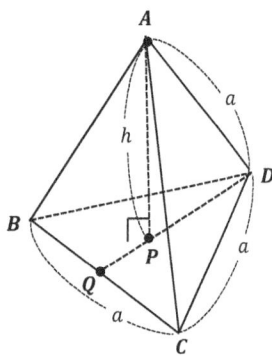

If all sides of a triangular pyramid have the same length a, then the altitude (height) is

$$h = \frac{\sqrt{6}\,a}{3}.$$

($\because$ Since the point P is the centroid of $\triangle BDC$, $DP : PQ = 2 : 1$.

Since $\triangle BDC$ is a regular triangle with sides of length a, the altitude DQ is $\dfrac{\sqrt{3}}{2}\,a$.

$\therefore\ DP = \dfrac{2}{3} \cdot DQ = \dfrac{2}{3} \cdot \dfrac{\sqrt{3}}{2}\,a = \dfrac{\sqrt{3}}{3}\,a.$

Since $\triangle APD$ is a right triangle,

$$a^2 = h^2 + DP^2 = h^2 + (\frac{\sqrt{3}}{3}\,a)^2 = h^2 + \frac{3}{9}a^2 = h^2 + \frac{1}{3}a^2.$$

That is, $h^2 = \dfrac{2}{3}a^2$. Therefore, $h = \sqrt{\dfrac{2}{3}a^2} = \dfrac{\sqrt{2}}{\sqrt{3}}\,a = \dfrac{\sqrt{6}\,a}{3}$.)

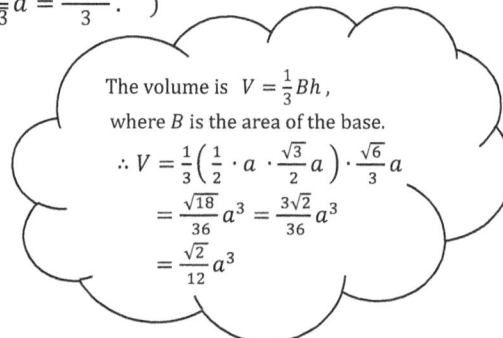

The volume is $V = \dfrac{1}{3}Bh$,

where B is the area of the base.

$$\therefore V = \frac{1}{3}\left(\frac{1}{2} \cdot a \cdot \frac{\sqrt{3}}{2}a\right) \cdot \frac{\sqrt{6}}{3}a$$
$$= \frac{\sqrt{18}}{36}a^3 = \frac{3\sqrt{2}}{36}a^3$$
$$= \frac{\sqrt{2}}{12}a^3$$

(5) Cones

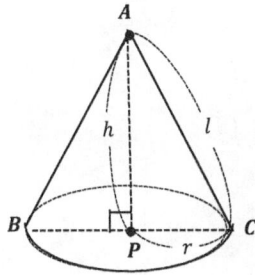

For a cone with radius r and slant height l, the altitude (height) is

$$h = \sqrt{l^2 - r^2}.$$

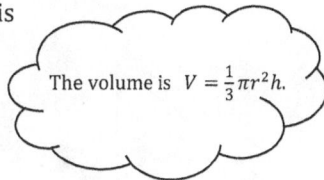

The volume is $V = \frac{1}{3}\pi r^2 h.$

($\because$ Since $\triangle APC$ is a right triangle, $l^2 = r^2 + h^2$; $h^2 = l^2 - r^2$.

Therefore, $h = \sqrt{l^2 - r^2}$.)

(6) Spheres

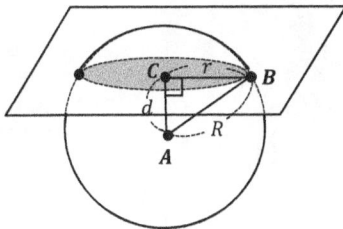

For a sphere of radius R, the horizontal cross sections are circular regions. If the cross section is at a distance d from the center of the sphere, and its radius is r, then

$$r = \sqrt{R^2 - d^2}.$$

($\because$ Since $\triangle ABC$ is a right triangle with $m\angle C = 90°$, $R^2 = d^2 + r^2$; $r^2 = R^2 - d^2$.

Therefore, $r = \sqrt{R^2 - d^2}$.)

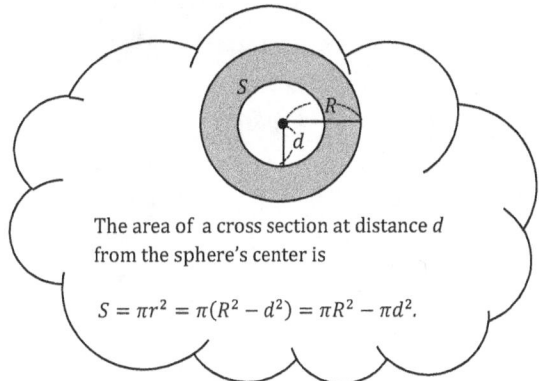

The area of a cross section at distance d from the sphere's center is

$$S = \pi r^2 = \pi(R^2 - d^2) = \pi R^2 - \pi d^2.$$

Exercises

#1 Find the value of x.

(1)

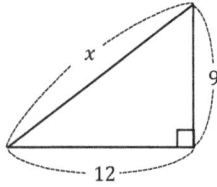

(2)

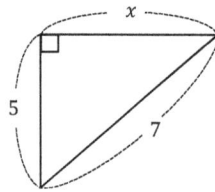

(3)

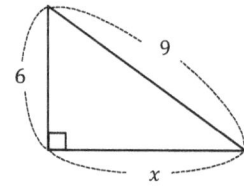

(4)

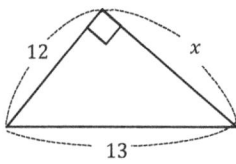

(5)

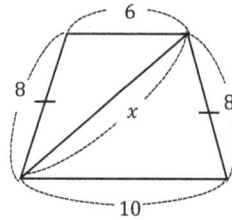

(6)

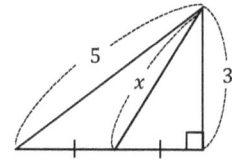

#2 Find the values of x and y.

(1)

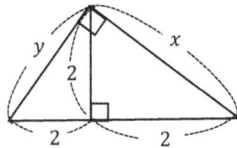

(2)

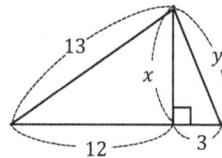

(3)

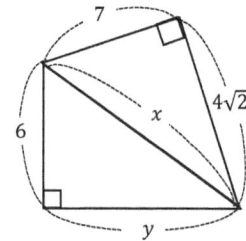

(4)

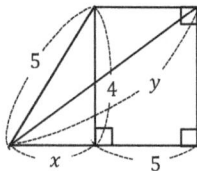

#3 Find the areas for the following :

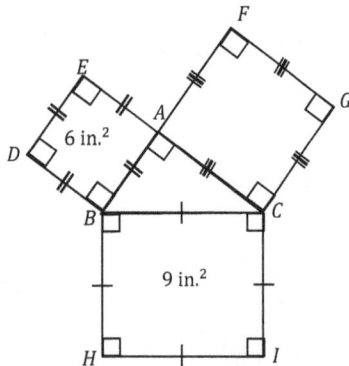

(1) The area of $\square BDEA$ is 6 in.^{2} and the area of $\square BCIH$ is 9 in.^{2} Find the area of $\square ACGF$.

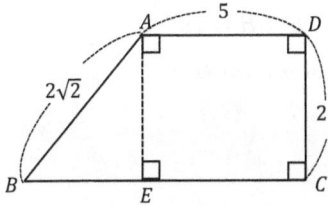

(2) Find the area of $\square ABCD$.

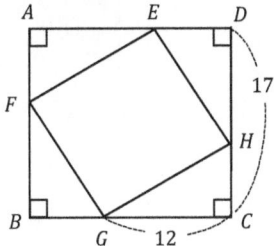

(3) All four right triangles are congruent.
Find the area of $\square EFGH$.

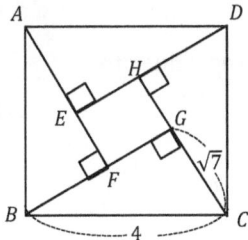

(4) Find the area of $\square EFGH$.

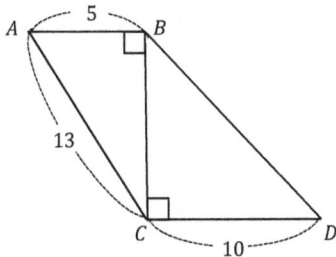

(5) Find the area of $\triangle BCD$.

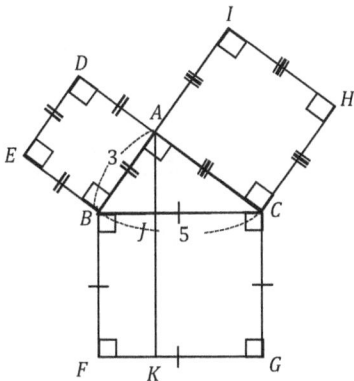

(6) Find the area of $\square JKGC$.

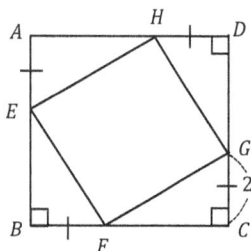

(7) The area of $\square EFGH$ is 18 in.^{2}

   Find the area of $\square ABCD$.

#4 Determine whether triangles with the following side lengths are right triangles.

(1) $2, 2, 4$

(2) $3, 4, 5$

(3) $1, 2, \sqrt{3}$

(4) $7, 9, 11$

(5) $\sqrt{2}, \sqrt{3}, \sqrt{5}$

(6) $5, 12, 13$

#5 The triangles $\triangle ABC$ are right triangles with $m\angle C = 90°$. Find the value of x.

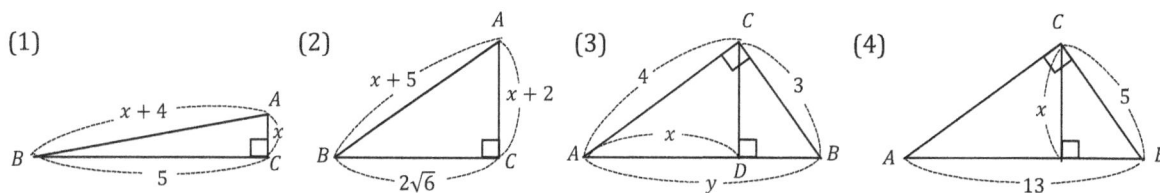

(1)

(2)

(3)

(4)

#6 Answer the questions for each figure.

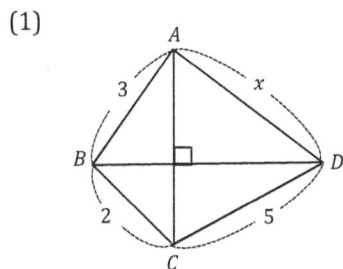

(1)

(2)

(3)

$\overline{AC} \perp \overline{BD}$. Find AD.

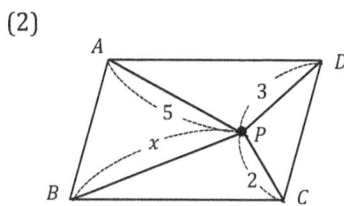

P is an interior point in $\square ABCD$.

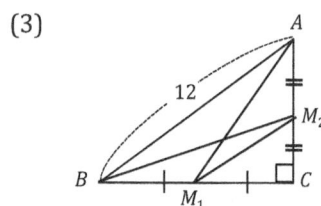

Find BP.

M_1 and M_2 are midpoints of $\overline{BC}$ and $\overline{AC}$, respectively.

Find $AM_1{}^2 + BM_2{}^2$.

(4)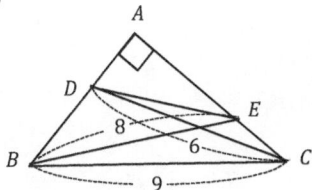

(5)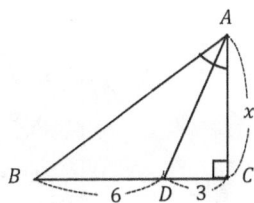

(6)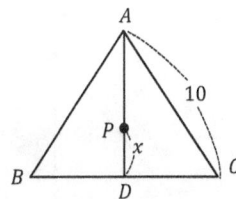

For a right triangle $\triangle ABC$ with $m\angle A = 90°$, $BE = 8, CD = 6$, and $BC = 9$. Find DE.

$\angle BAD \cong \angle DAC$
Find the value of x.

P is the centroid of a regular triangle $\triangle ABC$. Find the value of x.

#7 Find the value of x.

(1)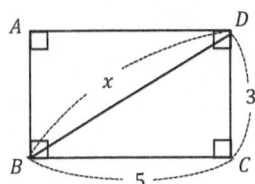

(2)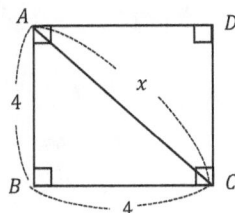

(3)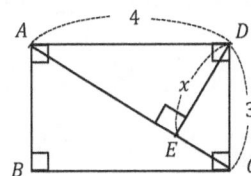

(4)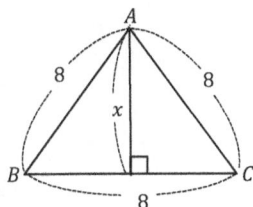

(5)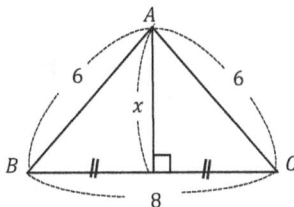

(6)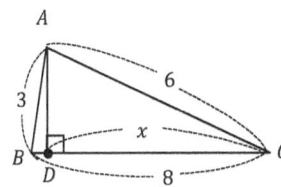

#8 Find the areas of $\triangle ABC$.

(1)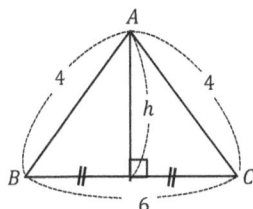

(2)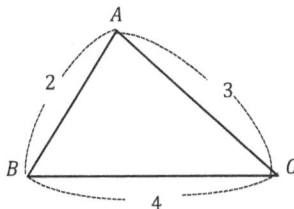

(3)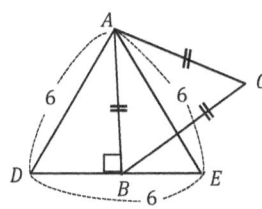

#9 Find the values of x and y.

(1)

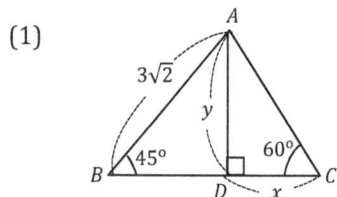

(2)

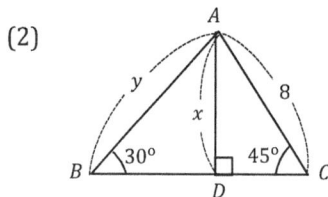

#10 Find the value of x.

(1)

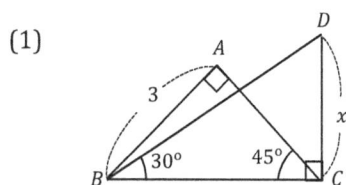

(2)

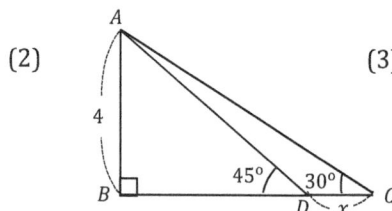

(3)

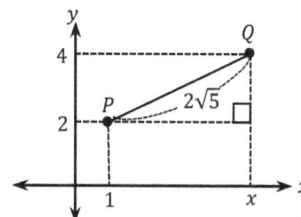

(4)

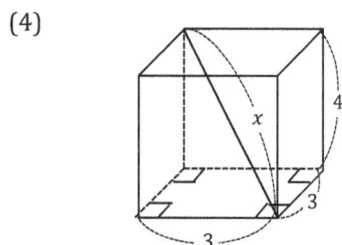

(5)

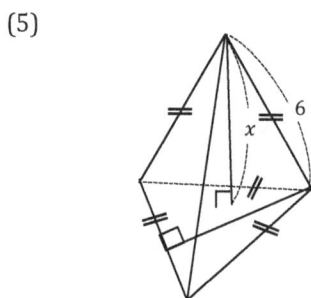

(6)

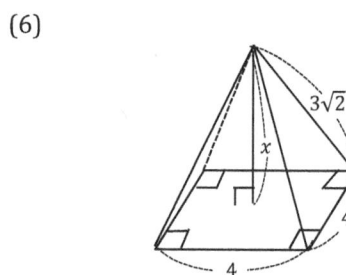

#11 Answer the questions for each figure.

(1)

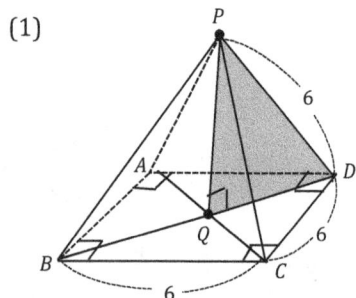

Find the area of $\triangle PQD$.

(2)

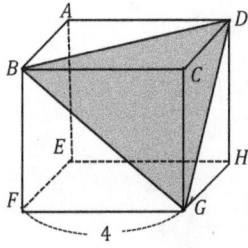

For a cube with sides of length 4, find the area of $\triangle BDG$.

(3)

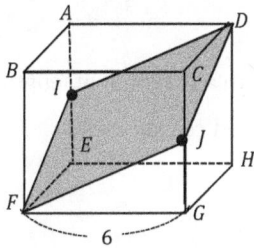

For a cube with sides of length 6, points I and J are midpoints of $\overline{AE}$ and $\overline{CG}$. Find the area of $\square IFJD$.

(4)

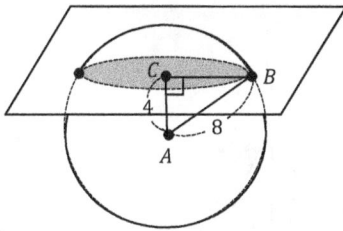

$AC = 4$, $AB = 8$.

Find the area of the cross section.

(5)

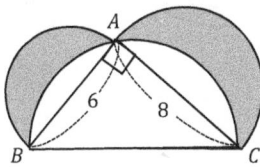

Find the area of the shaded regions.

(6)

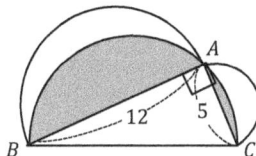

Find the area of the shaded regions.

(7)

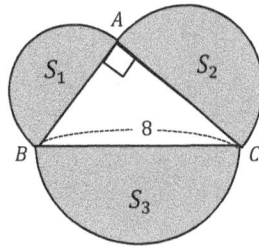

Find the area of the shaded regions.

(8)

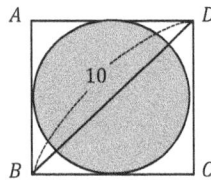

For a square $\square\ ABCD$ with a diagonal of length 10, find the area of the inscribed circle.

#12 Find the volume of each figure.

(1)

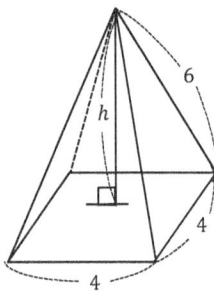

(2)

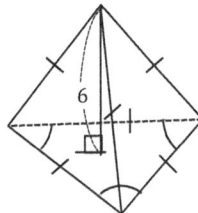

(3)

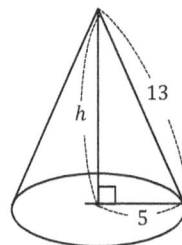

(4)

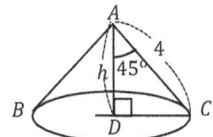

Trigonometric Ratios

CHAPTER 10

Chapter 10   Trigonometric Ratios

10-1   Basic Trigonometric Ratios

1. **Definition**
2. **The Trigonometric Ratios of the Special Triangles**

10-2   Numerical Trigonometry

1. **The Use of Tables**
2. **Measures of Sides**

    (1) **Right triangles**

    (2) **Triangles**

3. **Measures of an Altitude(Height)**
4. **Areas**

    (1) **The Area of a Triangle**

    (2) **The Area of a Quadrilateral**

10-3   Relationships among the Trigonometric Ratios

Chapter 10. Trigonometric Ratios

10-1   Basic Trigonometric Ratios

Trigonometry is the measurement of a triangle. Sine, cosine, and tangent are the main functions in trigonometry.

1. Definition

The three basic trigonometric ratios are sine, cosine, and tangent, briefly write sin, cos, and tan,

For a right triangle $\triangle ABC$ with $m\angle C = 90°$, find the trigonometric ratios of an acute angle $\angle A$.

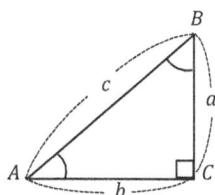

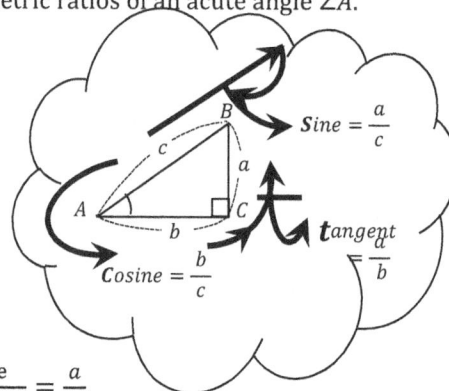

$$Sine = \frac{a}{c}$$
$$Cosine = \frac{b}{c}$$
$$tangent = \frac{a}{b}$$

The sine of $\angle A$ is

$$\sin A = \frac{\text{length of the side opposite the } \angle A}{\text{length of the hypotenuse}} = \frac{\text{opposite}}{\text{hypotenuse}} = \frac{a}{c}$$

The cosine of $\angle A$ is

$$\cos A = \frac{\text{length of the side adjacent to the } \angle A}{\text{length of the hypotenuse}} = \frac{\text{adjacent}}{\text{hypotenuse}} = \frac{b}{c}$$

The tangent of $\angle A$ is

$$\tan A = \frac{\text{length of the side opposite the } \angle A}{\text{length of the side adjacent to the } \angle A} = \frac{\text{opposite}}{\text{adjacent}} = \frac{a}{b}$$

If $m\angle A = \theta$, then we write $\sin\theta° = \frac{a}{c}$, $\cos\theta° = \frac{b}{c}$, and $\tan\theta° = \frac{a}{b}$.

Example

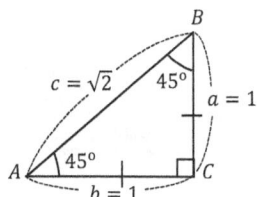

Consider a right $45°$-$45°$-$90°$ triangle.

Since trigonometric ratios do not depend on the size of the triangle, we can use any right triangle $\triangle ABC$ with $m\angle A = 45°$.

By the Pythagorean Theorem, $AC : BC : AB = 1 : 1 : \sqrt{2}$.

Taking $AC = BC = 1$,

$\sin \angle A = \sin 45° = \dfrac{a}{c} = \dfrac{1}{\sqrt{2}} = \dfrac{\sqrt{2}}{2}$

$\cos \angle A = \cos 45° = \dfrac{b}{c} = \dfrac{1}{\sqrt{2}} = \dfrac{\sqrt{2}}{2}$

$\tan \angle A = \tan 45° = \dfrac{a}{b} = \dfrac{1}{1} = 1$

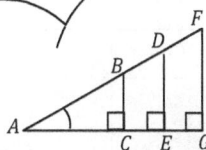

Since $\Delta ABC \sim \Delta ADE \sim \Delta AFG$, by AA Similarity Theorem,

$\sin A = \dfrac{BC}{AB} = \dfrac{DE}{AD} = \dfrac{FG}{AF}$

$\cos A = \dfrac{AC}{AB} = \dfrac{AE}{AD} = \dfrac{AG}{AF}$

$\tan A = \dfrac{BC}{AC} = \dfrac{DE}{AE} = \dfrac{FG}{AG}$.

The radios do not depend on the size of the triangle.

For $m \angle A = 30$,

$AC : BC : AB = \sqrt{3} : 1 : 2$, by the Pythagorean Theorem.

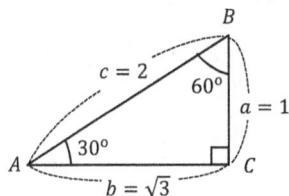

Taking $a = 1$, $b = \sqrt{3}$, and $c = 2$,

$\sin 30° = \dfrac{a}{c} = \dfrac{1}{2}$

$\cos 30° = \dfrac{b}{c} = \dfrac{\sqrt{3}}{2}$

$\tan 30° = \dfrac{a}{b} = \dfrac{1}{\sqrt{3}} = \dfrac{\sqrt{3}}{3}$

For $m \angle A = 60$,

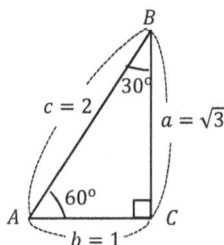

Taking $a = \sqrt{3}$, $b = 1$, and $c = 2$,

$\sin 60° = \dfrac{a}{c} = \dfrac{\sqrt{3}}{2}$

$\cos 60° = \dfrac{b}{c} = \dfrac{1}{2}$

$\tan 60° = \dfrac{a}{b} = \dfrac{\sqrt{3}}{1} = \sqrt{3}$

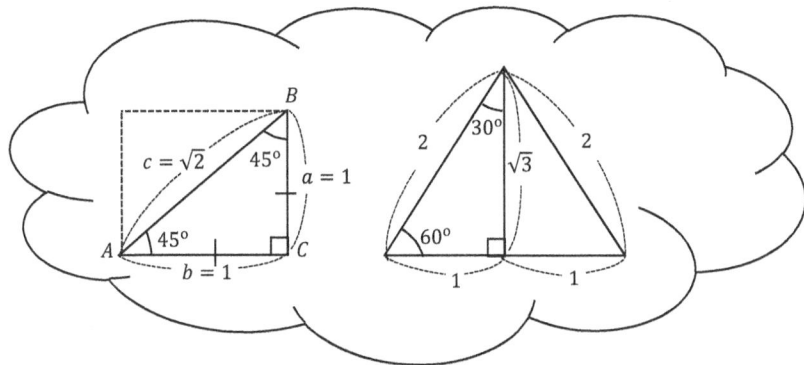

2. The Trigonometric Ratios of the Special Triangles

The trigonometric ratio $m \angle A$	30	45	60
$\sin A$	$\dfrac{1}{2}$	$\dfrac{\sqrt{2}}{2}$	$\dfrac{\sqrt{3}}{2}$
$\cos A$	$\dfrac{\sqrt{3}}{2}$	$\dfrac{\sqrt{2}}{2}$	$\dfrac{1}{2}$
$\tan A$	$\dfrac{\sqrt{3}}{3}$	1	$\sqrt{3}$

$$\sin 30° \left(= \tfrac{1}{2}\right) < \sin 45° \left(= \tfrac{\sqrt{2}}{2}\right) < \sin 60° (= \tfrac{\sqrt{3}}{2})$$

$$\cos 30° \left(= \tfrac{\sqrt{3}}{2}\right) > \cos 45° \left(= \tfrac{\sqrt{2}}{2}\right) > \cos 60° (= \tfrac{1}{2})$$

$\sin 30° = \cos 60°$
$\sin 60° = \cos 30°$
$\sin 45° = \cos 45°$

$\tan 30° = \dfrac{1}{\tan 60°}$
$\tan 45° = 1$

$y = ax + b$

$(0, y)$ $B(x, y)$

A C
$(0, 0)$ $(x, 0)$

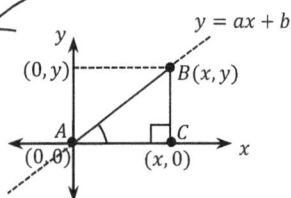

$$\tan A = \frac{BC}{AC} = \frac{y - 0}{x - 0} = \frac{\text{change in } y}{\text{change in } x}$$

Tangent is the ratio of the vertical change of a line to the horizontal change of a line

In other words, tangent is the slope of a line.

10-2 Numerical Trigonometry

1. The Use of Tables

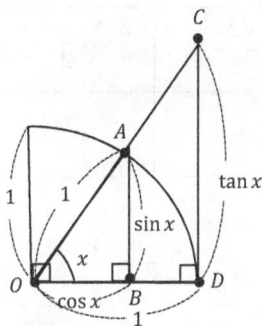

Consider a $\frac{1}{4}$ circle with a radius length of 1.

For triangles $\triangle AOB$ and $\triangle COD$ with the acute angle $\angle x$,

$\sin x = \frac{AB}{OA} = \frac{AB}{1} = AB$; Increasing function from 0 to 1.

$\cos x = \frac{OB}{OA} = \frac{OB}{1} = OB$; Decreasing function from 1 to 0.

$\tan x = \frac{CD}{OD} = \frac{CD}{1} = CD$; Increasing function from 0 to infinite ∞.

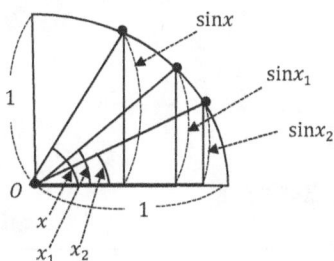

$x > x_1 > x_2 > x_3 > \cdots > 0$

$\Rightarrow \sin x > \sin x_1 > \sin x_2 > \sin x_3 > \cdots > 0$

$x \longrightarrow 0 \Rightarrow \sin x \longrightarrow 0 \quad (\sin 0° = 0)$

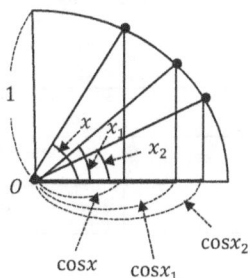

$x > x_1 > x_2 > x_3 > \cdots > 0$

$\Rightarrow \cos x < \cos x_1 < \cos x_2 < \cos x_3 < \cdots < 1$

$x \longrightarrow 0 \Rightarrow \cos x \longrightarrow 1 \quad (\cos 0° = 1)$

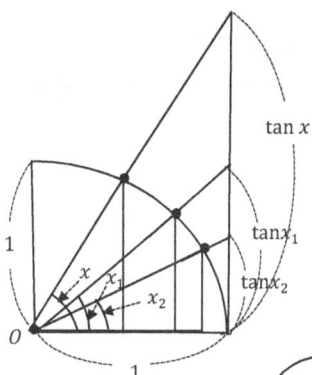

$x > x_1 > x_2 > x_3 > \cdots > 0$

$\Rightarrow \tan x > \tan x_1 > \tan x_2 > \tan x_3 > \cdots > 0$

$x \longrightarrow 0 \Rightarrow \tan x \longrightarrow 0 \quad (\tan 0° = 0)$

$x \longrightarrow 90 \Rightarrow \sin x \longrightarrow 1$
$(\sin 90° = 1)$
$x \longrightarrow 90 \Rightarrow \cos x \longrightarrow 0$
$(\cos 90° = 0)$
$x \longrightarrow 90 \Rightarrow \tan x \longrightarrow \infty$
$(\tan 90°$ does not defined. $)$

$\tan x = \frac{AB}{OB} = \frac{\sin x}{\cos x}$

The trigonometric ratio $\backslash m \angle A$	0	30	45	60	90
$\sin A$	0	$\frac{1}{2}$	$\frac{\sqrt{2}}{2}$	$\frac{\sqrt{3}}{2}$	1
$\cos A$	1	$\frac{\sqrt{3}}{2}$	$\frac{\sqrt{2}}{2}$	$\frac{1}{2}$	0
$\tan A$	0	$\frac{\sqrt{3}}{3}$	1	$\sqrt{3}$	

Rounding $\sqrt{2}$ and $\sqrt{3}$ to the thousandths place, we find $\sqrt{2} = 1.414$, $\sqrt{3} = 1.732$

So, $\sin 30° = \frac{1}{2} = .500$, $\cos 30° = \frac{\sqrt{3}}{2} = \frac{1.732}{2} = .866$, and $\tan 30° = \frac{\sqrt{3}}{3} = \frac{1.732}{3} = .577$

With this strategy, we can calculate trigonometric ratios.

Table of Trigonometric Ratios

x	$\sin x$	$\cos x$	$\tan x$
$\vdots$	$\vdots$	$\vdots$	$\vdots$
30°	.500	.866	.577
31°	.515	.857	.601
32°	.530	.848	.625
$\vdots$	$\vdots$	$\vdots$	$\vdots$
45°	.707	.707	1.000
$\vdots$	$\vdots$	$\vdots$	$\vdots$
60°	.866	.500	1.732
$\vdots$	$\vdots$	$\vdots$	$\vdots$
90°	1.000	.000	

Example   Suppose $AB = 15$ ft and $x = 31$.

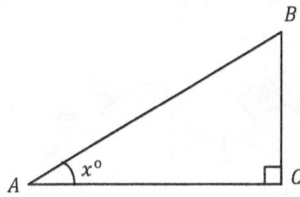

To measure BC, use the table of trigonometric ratios.

Since $\sin x° = \dfrac{BC}{AB}$, $BC = AB \sin x°$.

Therefore, $BC = 15 \times \sin 31° = 15 \times .515 = 7.725$ ft

2. Measures of Sides

(1) Right triangles

For a right triangle $\triangle ABC$ with $m \angle C = 90°$,

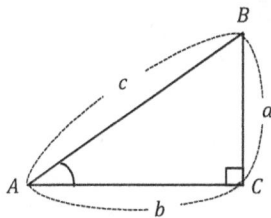

1) Given $m \angle A$ and a length c of AB

   $\Rightarrow a = c \sin A$ or $b = c \cos A$

2) Given $m \angle A$ and a length b of AC

   $\Rightarrow a = b \tan A$ or $c = \dfrac{b}{\cos A}$

3) Given $m \angle A$ and a length a of BC

   $\Rightarrow c = \dfrac{a}{\sin A}$ or $b = \dfrac{a}{\tan A}$

$\sin A = \dfrac{a}{c} \Rightarrow a = c \sin A$
$\cos A = \dfrac{b}{c} \Rightarrow b = c \cos A$
$\tan A = \dfrac{a}{b} \Rightarrow a = b \tan A$

(2) Triangles

For a triangle $\triangle ABC$,

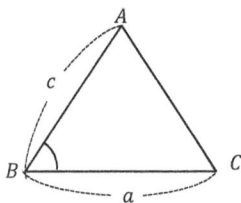

1) Given the lengths of two sides and the measure of the included angle,

   $AC = \sqrt{(c \sin B)^2 + (a - c \cos B)^2}$ by the Pythagorean Theorem.

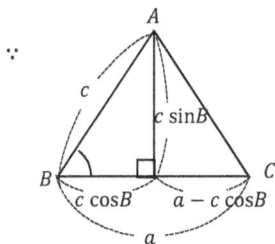

$\therefore$

$c \sin B$

$c \cos B$ $a - c \cos B$

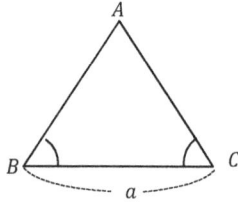

2) Given the measures of two angles and the length of the included side,

$$AC = \frac{a \sin B}{\sin A}, \quad AB = \frac{a \sin C}{\sin A}$$

$\therefore$

$$\sin B = \frac{CD}{a}$$
$$CD = a \sin B$$

$$\sin A = \frac{a \sin B}{AC}$$
$$AC = \frac{a \sin B}{\sin A}$$

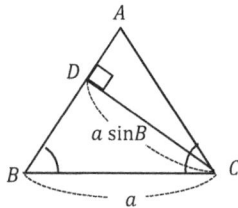

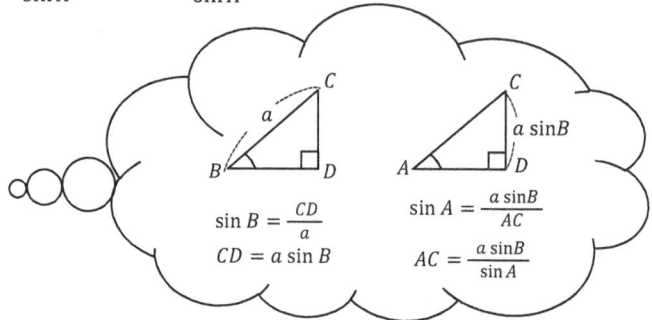

3. Measures of an Altitude(Height)

Given the measures of two angles and the length of the included side, we can measure the altitude of a triangle.

(1) For a triangle with acute angles, $h = \dfrac{a}{\tan x + \tan y}$

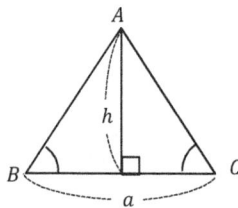

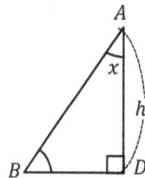

$$\tan x = \frac{BD}{AD}$$
$$BD = AD \tan x$$
$$\therefore BD = h \tan x$$

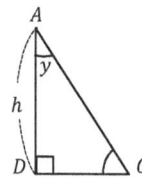

$$\tan y = \frac{DC}{AD}$$
$$DC = AD \tan y$$
$$\therefore DC = h \tan y$$

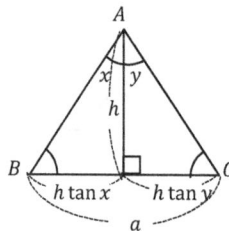

$$a = h \tan x + h \tan y$$
$$= h (\tan x + \tan y)$$
$$\therefore h = \frac{a}{\tan x + \tan y}$$

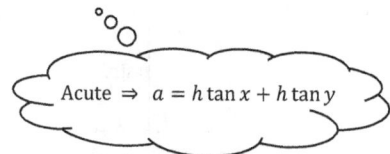

Acute $\Rightarrow a = h \tan x + h \tan y$

(2) For a triangle with an obtuse angle, $h = \dfrac{a}{\tan x - \tan y}$

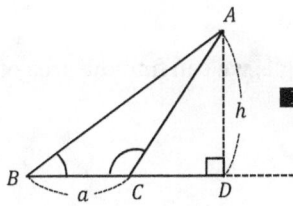

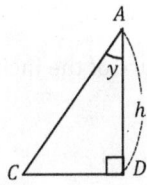

$$\tan y = \frac{CD}{AD}$$
$$CD = AD \tan y$$
$$\therefore CD = h \tan y$$

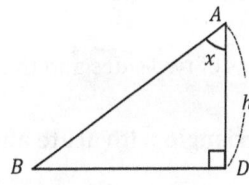

$$\tan x = \frac{BD}{AD}$$
$$BD = AD \tan x$$
$$\therefore BD = h \tan x$$

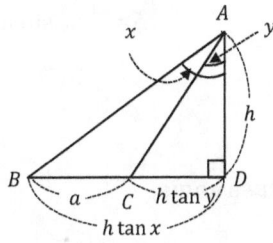

$$a + h \tan y = h \tan x$$
$$a = h \tan x - h \tan y$$
$$= h (\tan x - \tan y)$$
$$\therefore h = \frac{a}{\tan x - \tan y}$$

Obtuse $\Rightarrow a = h \tan x - h \tan y$

4. Areas

(1) The Area of a Triangle

Given lengths of two sides and the measure of the included angle, we can find the area of a triangle.

1) For a triangle with acute angles,

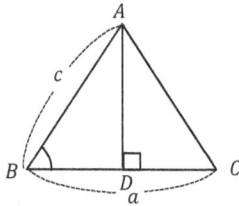

Since $AD = c \sin B$,

the area of $\triangle ABC$ is

$$S = \frac{1}{2} ac \sin B.$$

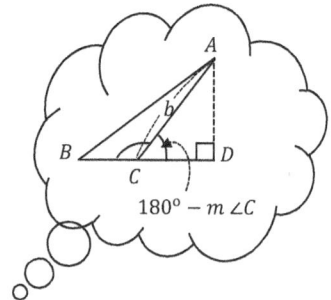

2) For a triangle with an obtuse angle,

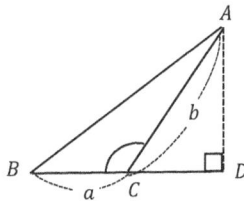

Since $AD = b \sin (180° - m \angle C)$,

the area of $\triangle ABC$ is

$$S = \frac{1}{2} ab \sin (180° - m \angle C).$$

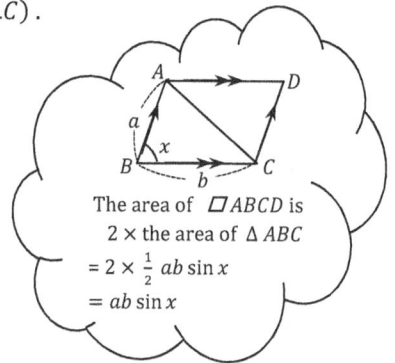

The area of $\square ABCD$ is
$2 \times$ the area of $\triangle ABC$
$= 2 \times \frac{1}{2} ab \sin x$
$= ab \sin x$

> For a parallelogram,
> any two opposite sides are congruent and
> any two opposite angles are congruent.

(2) The Area of a Quadrilateral

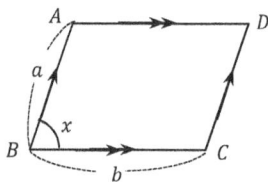

1) For a parallelogram $\square ABCD$ with length a and b of two sides and the measure of the included acute angle $\angle x$,

the area of $\square ABCD$ is $S = ab \sin x$.

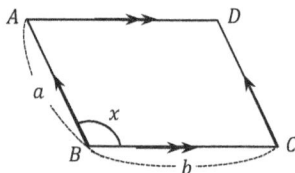

If $\angle B$ is obtuse, then $S = ab \sin (180° - x)$.

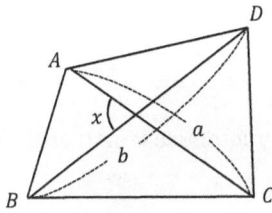

2) For a quadrilateral $\Box\,ABCD$ with the length of two diagonals and the measure of the acute angle $\angle\,x$ formed by the two diagonals,

the area of $\Box\,ABCD$ is $S = \frac{1}{2}ab\sin x$.

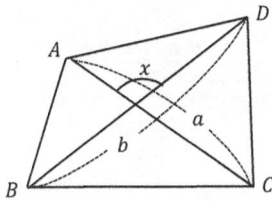

If $\angle\,x$ is obtuse, then $S = \frac{1}{2}ab\sin(180° - x)$.

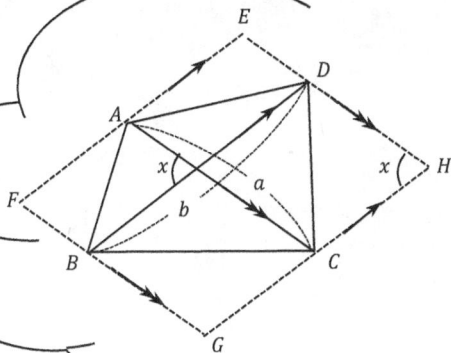

Consider a parallelogram $\Box\,EFGH$ with $\overline{EF}\,/\!/\,\overline{DB}\,/\!/\,\overline{HG}$ and $\overline{EH}\,/\!/\,\overline{AC}\,/\!/\,\overline{FG}$.

The area of $\Box\,ABCD$ is

$\frac{1}{2}\times$ the area of $\Box EFGH$

$= \frac{1}{2}\times ab\sin x$

$= \frac{1}{2}ab\sin x$

10-3 Relationships among the Trigonometric Ratios

For any angle, the square of the sine plus the square of the cosine is 1.

Applying the Pythagorean Theorem to a right triangle with the hypotenuse length of 1, the *Pythagorean identity* is written

$$\sin^2 A + \cos^2 A = 1 \quad \text{for any angle } A.$$

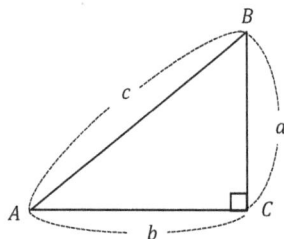

By the Pythagorean Theorem, $c^2 = a^2 + b^2$.

Dividing by c^2, $1 = \left(\frac{a}{c}\right)^2 + \left(\frac{b}{c}\right)^2$.

Since $\sin A = \frac{a}{c}$ and $\cos A = \frac{b}{c}$,

$$\sin^2 A + \cos^2 A = 1.$$

> The square of the sine of $\angle A$ is $(\sin A)^2 = \sin^2 A$.

This gives, for any angle $\angle A$,

$$\boxed{\sin^2 A + \sin^2 B = 1.}$$

Note : For any angle, $\angle A$, $\tan A = \frac{\sin A}{\cos A}$.

For the right triangle $\triangle ABC$ with $\angle C = 90^\circ$, $\sin A = \frac{a}{c} = \cos B$ *and* $\cos A = \frac{b}{c} = \sin B$.

This means if $\angle A$ and $\angle B$ are complementary ($m \angle A + m \angle B = 90^\circ$), then

$\sin A = \cos B$ *and* $\cos A = \sin B$.

$\sin x^\circ = \cos(90 - x)^\circ$ *and* $\cos x^\circ = \sin(90 - x)^\circ$

> The sine of an angle is the cosine of its complement.

Exercises

#1 Find the trigonometric ratios of $\angle x$.

(1)

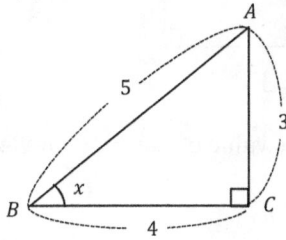

(2)

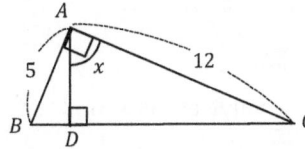

#2 Find the value of the following :

(1) $\sin 30° + \cos 30°$

(2) $\sin 45° + \cos 60°$

(3) $2 \tan 30° \times \sin 60° \times \cos 45°$

(4) $\sin^2 60 + \cos^2 60$

#3 Find the values of x and y for the following right triangles $\triangle ABC$.

(1)

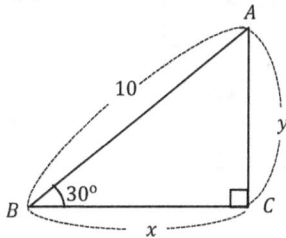

(2)

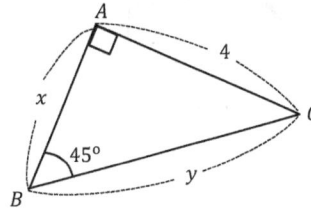

(3)

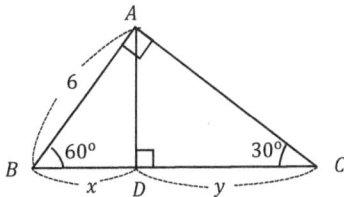

(4)

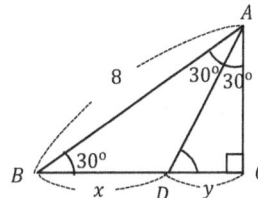

#4 For right triangles $\triangle ABC$ with $m\angle C = 90°$, answer the following :

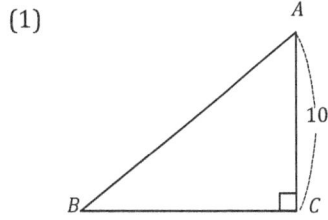

(1)

$\sin B = \dfrac{5}{13}$

Find the value of $\sin A$ and $\tan A$.

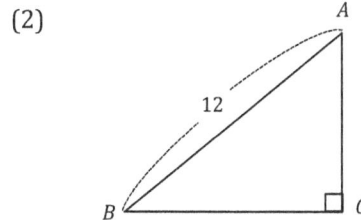

(2)

$\sin A = \dfrac{2}{3}$

Find the value of $\cos B \times \tan B$.

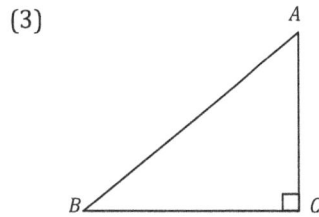

(3)

$m\angle A : m\angle B : m\angle C = 1 : 2 : 3$

Find the value of $\sin B \times \cos B \times \tan B$.

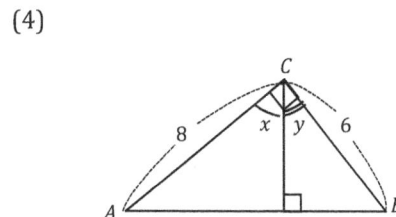

(4)

Find the value of $\sin x + \cos y$.

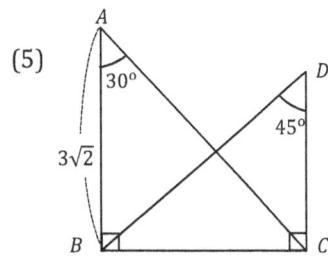

(5)

$m\angle C = m\angle B = 90°$

Find BD.

(6)

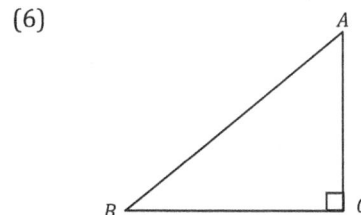

$\sin A = \dfrac{\sqrt{7}}{4}$

Find $\tan B$.

#5 Find the value of x for the following triangles:

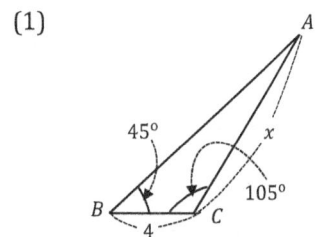

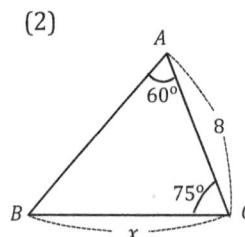

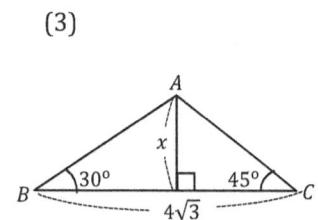

(1)

(2)

(3)

(4)

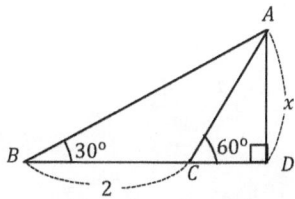

(5)

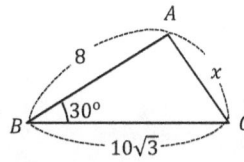

#6 Find the areas of the following :

(1)

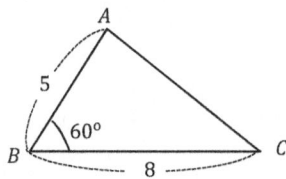

(2)

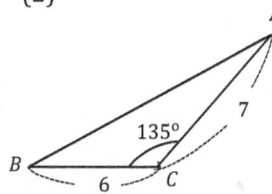

(3)

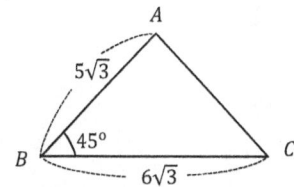

(4)

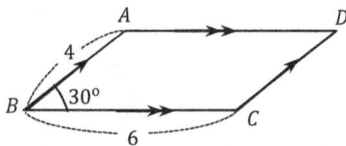

(5)

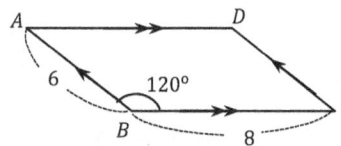

(6)

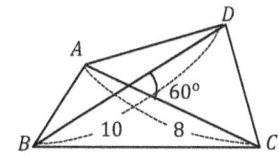

(7)

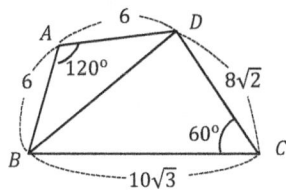

(8)

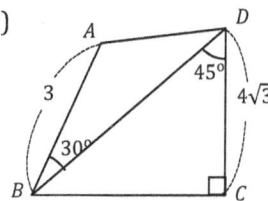

Solutions Manual

Geometry

Solutions for Chapter1

#1 **Find the distance for the following from the figure below :**

   (1) **The distance from point A to point B.** ; $AB = 5$

   (2) **The distance from point B to point D.** ; $BD = 2 + 3 = 5$

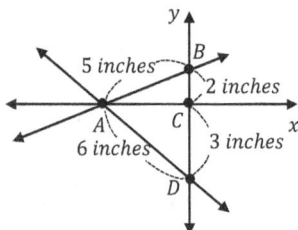

#2 **Let points M and N be the midpoints of segments $\overline{AB}$ and $\overline{MB}$, respectively.**

   (1) **Determine whether the following expressions are the true or false.**

   1) $AM = BM$; true

   2) $AM = \frac{1}{2}AB$; true

   3) $NB = \frac{1}{3}AB$; false ($\because NB = \frac{1}{4}AB$)

   4) $AB = 2MN$; false ($\because AB = 4MN$)

   5) $MN = \frac{1}{2}AM$; true

   (2) **If the distance AB is 20 inches, find the distance MN and the distance AN.**

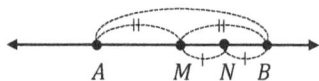

Since M is the midpoint of $\overline{AB}$, $AM = MB = \frac{20}{2} = 10$ inches.

Since N is the midpoint of $\overline{MB}$, $MN = NB = \frac{10}{2} = 5$ inches

Since $\overline{AN} = \overline{AM} + \overline{MN}$, the distance AN is 10+5=15 inches.

#3 **Determine whether each statement is true or false.**

   (1) **If two lines are parallel to another line, then the two lines are parallel to each other.** ; true

   (2) **Two lines on a plane always intersect at one point.** ; false

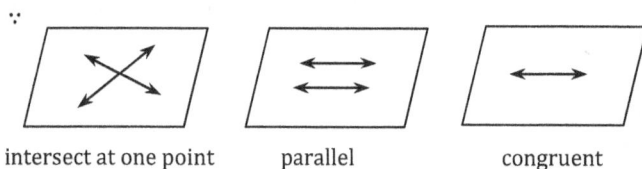

intersect at one point   parallel   congruent

(3) **If two lines are not intersecting, then the lines are not on a plane.** ; false

∵

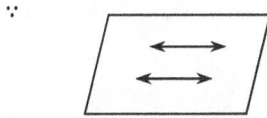

(4) **There is always a plane containing two distinct lines.** ; false

∵

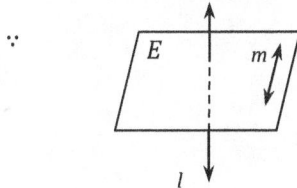

l and *m* are two distinct lines. But *l* does not contained in plane *E*.

(5) **Two parallel lines do not have intersection points.** ; true

(6) **If two distinct lines intersect, their intersection contains only one point.** ; true

(7) **If two points of a line lie in a plane, then the line lies in the same plane.** ; true

(8) **If two different planes intersect, their intersection is a point.** ; false

(∵ Their intersection is a line.)

#4 Using the figure, find the measure of each angle.

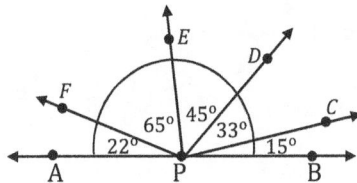

(1) $\angle BPC$; $m\angle BPC = 15°$

(2) $\angle CPE$; $m\angle CPE\ = 33° + 45° = 78°$

(3) $\angle CPF$; $m\angle CPF = 33° + 45° + 65° = 143°$

(4) $\angle APE$; $m\angle APE = 22° + 65° = 87°$

#5 Using the figure, identify each angle as acute, right, or obtuse.

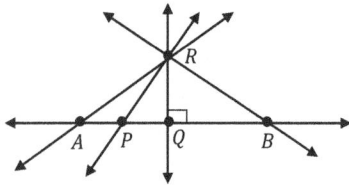

    (1) $\angle RAB$; acute

    (2) $\angle PRQ$; acute

    (3) $\angle RQP$; right

    (4) $\angle APR$; obtuse

#6 Determine the measure of the complement of an angle with a measure of :

    (1) 36° ; $90° - 36° = 54°$

    (2) 75° ; $90° - 75° = 15°$

    (3) 24.5° ; $90° - 24.5° = 65.5°$

    (4) 45° + n° ; $90° - (45° + n°) = 45° - n°$

    (5) 90° - n° ; $90° - (90° - n°) = n°$

#7 Using the figure, answer the following :

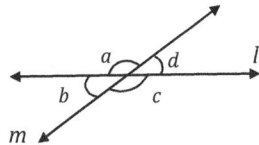

    (1) If $m\angle a = 120°$, what is $m\angle b$?

    Since $\angle a$ is the supplement of $\angle b$, $m\angle b = 180° - 120° = 60°$.

    (2) If $m\angle d = 43°$, what are $m\angle a$, $m\angle b$, and $m\angle c$?

    $m\angle a = 180° - 43° = 137°$

    Since $\angle b$ and $\angle d$ are vertical angles, $m\angle b = m\angle d = 43°$.

    Since $\angle a$ and $\angle c$ are vertical angles, $m\angle a = m\angle c = 137°$.

    Therefore, $m\angle a = 137°$, $m\angle b = 43°$, and $m\angle c = 137°$.

#8 For the following figures, find the measures of $\angle a$:

(1)

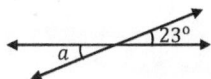

(2)

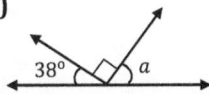

(3)

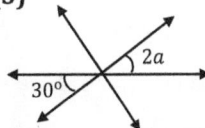

   (1) $m \angle a = 23°$, by vertical angle.

   (2) Since $38° + 90° + m \angle a = 180°$, $m \angle a = 90° - 38° = 52°$

   (3) Since $m \angle 2a = 2(m \angle a) = 30°$, by vertical angles, $m \angle a = 15°$

(4)

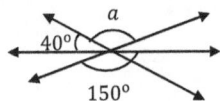

(5)

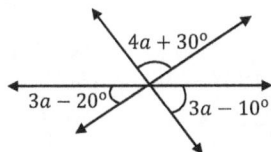

   (4) Since $150° = 40° + m \angle a$, $m \angle a = 150° - 40° = 110°$

   (5) $180° = m \angle(3a - 20°) + m \angle(4a + 30°) + m \angle(3a - 10°)$

            $= m \angle 3a + m \angle 4a + m \angle 3a = 3(m \angle a) + 4(m \angle a) + 3(m \angle a) = 10(m \angle a)$

      $\therefore$ $m \angle a = 18°$

#9 Lines l and m are parallel. Using the following figures, find the measures of the angles.

  (1)

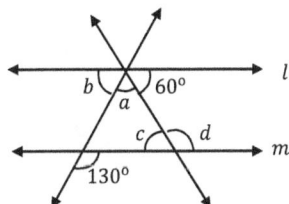

    ① **$\angle a$** ; by corresponding angles, $m \angle a + 60° = 130°$. So, $m \angle a = 70°$

    ② **$\angle b$** ; by vertical angles, $m \angle b = 50°$

    ③ **$\angle c$** ; by alternate interior angles, $m \angle c = 60°$

    ④ **$\angle d$** ; Since $\angle d$ is a supplement of $\angle c$, $m \angle d = 180° - m \angle c = 120°$

(2)

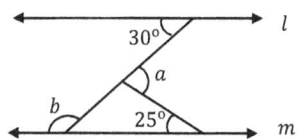

① $\angle a$ ② $\angle b$

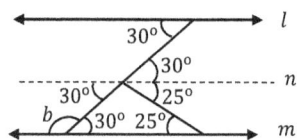

Let $n /\!/ l$. Then, $n /\!/ m$.

$\therefore$ $m\angle a = 30° + 25° = 55°$ and $m\angle b = 180° - 30° = 150°$

Solutions for Chapter 2

#1 Given the following lengths of sides, determine whether or not each example can be a triangle.

(1) 4, 5, 6 ; triangle is defined. ($\because 6 < 4 + 5$)

(2) 3, 4, 8 ; triangle is not defined. ($\because 8 \nless 3 + 4$)

(3) 5, 6, 11 ; triangle is not defined. ($\because 11 \nless 5 + 6$)

(4) 7, 7, 7 ; triangle is defined. ($\because 7 < 7 + 7$)

#2 Given the following side lengths, what kinds of triangles are each of the following?

(1) 6, 8, 6 ; isosceles

(2) 5, 5, 5 ; equilateral

(3) 4, 5, 6 ; scalene

(4) 3, 5, 7 ; scalene

#3 Identify which correspondence (SSS, SAS, ASA, HL, SAA, or none) must be applied for the following congruent triangles :

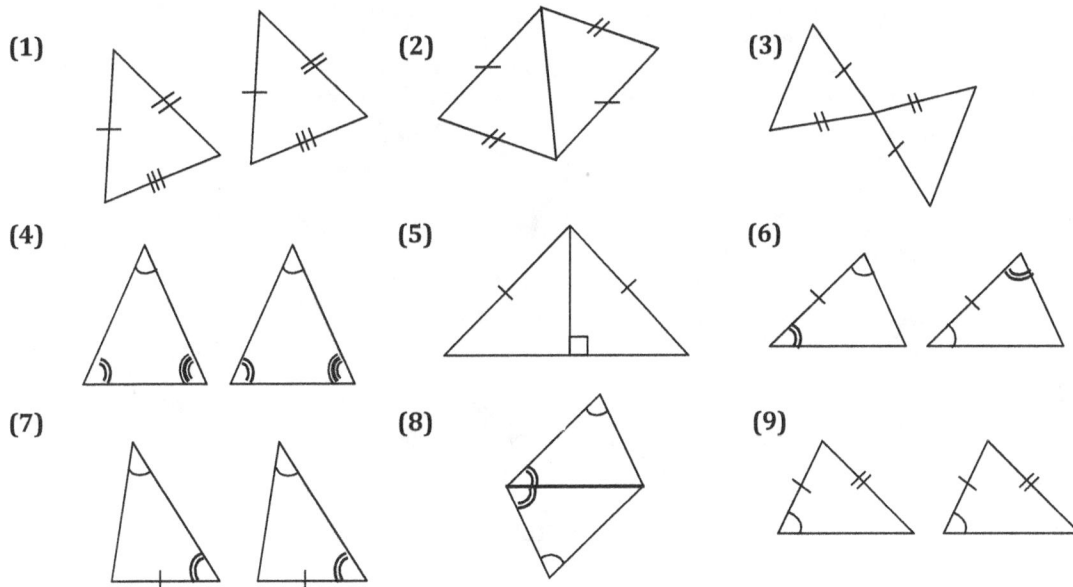

(1)   **(2)**   **(3)**

(4)   **(5)**   **(6)**

(7)   **(8)**   **(9)**

(1) SSS (2) SSS (3) SAS (4) None (5) HL (6) ASA (7) SAA (8) SAA (9) None

#4 Find the measure of angle $\angle a$ for the following :

(1)

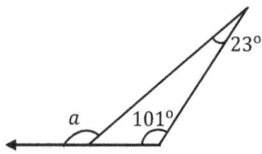

(2)

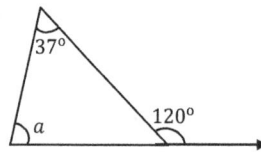

(3)

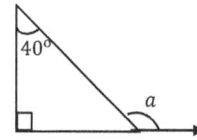

(4)

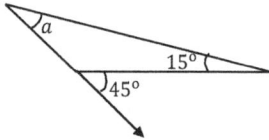

(5)

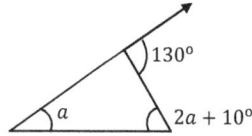

(6)

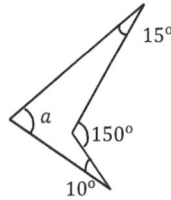

(7)

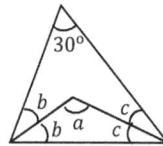

(1) $m \angle a = 23° + 101° = 124°$

(2) $120° = 37° + m \angle a$ $\therefore m \angle a = 83°$

(3) $m \angle a = 40° + 90° = 130°$

(4) $45° = m \angle a + 15°$ $\therefore m \angle a = 30°$

(5) $m \angle a + m \angle(2a + 10°) = 130°$; $3(m \angle a) + 10° = 130°$ $\therefore m \angle a = \frac{120°}{3} = 40°$

(6)

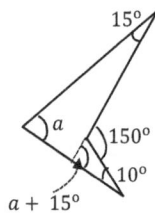

$150° = (m \angle a + 15°) + 10° = m \angle a + 25°$

$\therefore m \angle a = 125°$

(7)   Since $m \angle 2b + m \angle 2c + 30° = 180°$, $m \angle 2b + m \angle 2c = 150°$; $2(m \angle b + m \angle c) = 150°$

$\therefore m \angle b + m \angle c = 75°$

Since $m \angle a + m \angle b + m \angle c = 180°$, $m \angle a = 180° - 75° = 105°$

#5 Find the measure of $\angle a$.

(1)

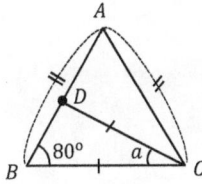

$\overline{AB} \cong \overline{AC}$, $\overline{BC} \cong \overline{CD}$, and

$m \angle B = 80^\circ$

Since $\overline{AB} \cong \overline{AC}$, $\triangle ABC$ is isosceles, So, $m \angle B = m \angle C$ ∴ $m \angle ACD = 80^\circ - m \angle a$.

Since $\overline{BC} \cong \overline{CD}$, $m \angle CDB = 80^\circ$. ∴ $m \angle ADC = 100^\circ$

Since $\angle ADC$ is an exterior angle of $\triangle DBC$, $m \angle ADC = m \angle DBC + m \angle a$.

So, $100^\circ = 80^\circ + m \angle a$. ∴ $m \angle a = 20^\circ$.

(2)

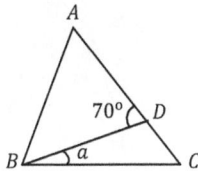

$\overline{AB} \cong \overline{AC} \cong \overline{BD}$ and

$m \angle ADB = 70^\circ$

Since $\overline{AB} \cong \overline{BD}$, $m \angle BAD = 70^\circ$.

Since $\overline{AB} \cong \overline{AC}$, $m \angle ABC = m \angle ACB = \frac{180^\circ - 70^\circ}{2} = 55^\circ$.

Since $70^\circ = m \angle a + m \angle DCB$ and $m \angle DCB = m \angle ACB = 55^\circ$, $m \angle a = 15^\circ$

(3)

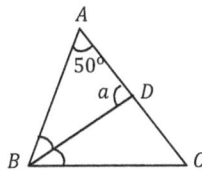

$\overline{AB} \cong \overline{AC}$, $m \angle A = 50^\circ$ and

$\angle ABD \cong \angle DBC$

Since $\overline{AB} \cong \overline{AC}$, $\angle B \cong \angle C$. ∴ $m \angle B = m \angle C = \frac{180^\circ - 50^\circ}{2} = 65^\circ$.

Since $\angle ABD \cong \angle DBC$ and $\angle a$ is an exterior angle of $\triangle DBC$,

$m \angle a = m \angle DBC + m \angle C = \frac{65^\circ}{2} + 65^\circ = 97.5^\circ$

(4)

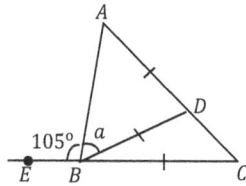

$\overline{AD} \cong \overline{BD} \cong \overline{BC}$ and

$m \angle ABE = 105°$

Since $\overline{AD} \cong \overline{BD}$, $\angle DAB \cong \angle a$.

Since $\angle CDB$ is an exterior angle of $\triangle ABD$, $m \angle CDB = 2(m \angle a)$

Since $\overline{BD} \cong \overline{BC}$, $\angle C \cong \angle CDB$

$\therefore m \angle C = 2 (m \angle a)$ $\therefore m \angle DBC = 180° - 4 (m \angle a)$

Since $m \angle DBC + m \angle a = 180° - 105° = 75°$, $m \angle DBC = 75° - m \angle a$

So, $75° - m \angle a = 180° - 4 (m \angle a)$; $3 (m \angle a) = 105°$ Therefore, $m \angle a = 35°$

(5)

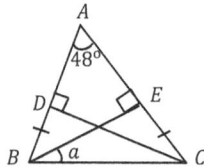

$m \angle A = 48°$ and

$\overline{DB} \cong \overline{EC}$

Since $\triangle DBC \cong \triangle EBD$, by RHS(HL) Postulate, $\angle B \cong \angle C$

Since $m \angle A = 48°$, $m \angle B = m \angle C = \dfrac{180° - 48°}{2} = \dfrac{132°}{2} = 66°$

From $\triangle EBC$, $m \angle a = 180° - (90° + 66°) = 24°$

(6)

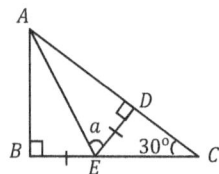

$m \angle B = 90°$, $m \angle C = 30°, \overline{AC} \perp$

$\overline{DE}$, $\overline{BE} \cong \overline{DE}$

From $\triangle ABC$, $m \angle A = 180° - (90° + 30°) = 60°$

Since $\triangle ABE \cong \triangle ADE$, by HL Theorem (RHS correspondence), $m \angle BAE = m \angle EAD = 30°$

From $\triangle AED$, $m \angle a = 180° - (30° + 90°) = 60°$

#6 $\triangle ABC$ is a right triangle with $m \angle C = 90°$. Let $m \angle B = 30°$ and M be the midpoint of $\overline{AB}$.

Find $m \angle ACM$ and $m \angle BMC$.

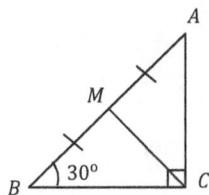

From $\triangle ABC$, $m \angle A = 180° - (90° + 30°) = 60°$

Since M is the midpoint of $\overline{AB}$, $\overline{AM} \cong \overline{MB} \cong \overline{MC}$

So, $\triangle AMC$ is isosceles. $\therefore m \angle A = m \angle ACM$ $\therefore m \angle ACM = 60°$

From $\triangle AMC$, $m \angle AMC = 180° - (60° + 60°) = 60°$

$\therefore m \angle BMC = 180° - 60° = 120°$

#7 How many diagonals has a polygon with 6 sides ; 12 sides ?

6 sides $\Rightarrow \frac{n(n-3)}{2} = \frac{6(6-3)}{2} = 9,$ 12 sides $\Rightarrow \frac{n(n-3)}{2} = \frac{12(12-3)}{2} = 54$

#8 How many triangles can be formed using the vertices of a hexagon ; decagon ?

$(n - 2)$ triangles

$\therefore$ Hexagon $\Rightarrow 6 - 2 = 4$ triangles and Decagon $\Rightarrow 10 - 2 = 8$ triangles

#9 Find the sum of the measures of the interior angles of a convex pentagon ; of a convex octagon.

Convex pentagon $\Rightarrow (n - 2) \times 180° = (5 - 2) \times 180° = 540°$

Convex octagon $\Rightarrow (n - 2) \times 180° = (8 - 2) \times 180° = 1080°$

#10 Find the number of sides of a convex polygon if the sum of the measures of its angles is 900 ; 1620 ; 2340.

$(n - 2) \times 180° = 900 \therefore n = 7$

$(n - 2) \times 180° = 1620 \therefore n = 11$

$(n - 2) \times 180° = 2340 \therefore n = 15$

#11 Find the measures of each interior and exterior angle of a regular pentagon and a regular octagon.

Interior angle : $\dfrac{(n-2)180^0}{n}$

Exterior angle : $\dfrac{360^0}{n}$

Regular pentagon $\Rightarrow$ Interior angle is $\dfrac{(5-2)180^0}{5} = 108^\circ$ and exterior angle is $\dfrac{360^0}{5} = 72^\circ$.

Regular octagon $\Rightarrow$ Interior angle is $\dfrac{(8-2)180^0}{8} = 135^\circ$ and exterior angle is $\dfrac{360^0}{8} = 45^\circ$.

#12 Find the measure of angle $\angle a$ of the following :

(1)

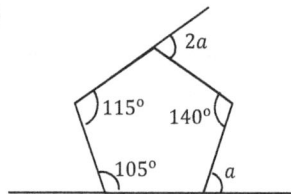

$n = 5 \Rightarrow (n - 2) \times 180^\circ = 540^\circ$

$\therefore (180^\circ - 2\,m \angle a) + 115^\circ + 105^\circ + (180^\circ - m \angle a) + 140^\circ = 540^\circ$

$720^\circ - 3\,m \angle a = 540^\circ \ ; \ 3\,m \angle a = 180^\circ \quad \therefore \ m \angle a = 60^\circ$

(2)

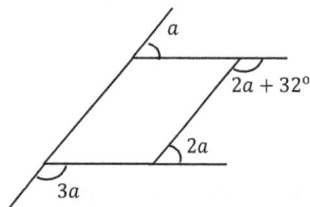

$8\,m \angle a + 32^\circ = 360^\circ \ ; \ 8\,m \angle a = 328^\circ \ \therefore \ m \angle a = \dfrac{328^0}{8} = 41^\circ$

Solutions for Chapter 3

#1 Determine whether the following statements are true or false.

    (1) A square is a rectangle. ; true

    (2) A rhombus is a square. ; false

    (3) A rectangle is a trapezoid. ; true

    (4) A square is an isosceles trapezoid. ; true

    (5) A square is a parallelogram. ; true

    (6) A rhombus is an isosceles trapezoid. ; false

    (7) A rectangle is an isosceles trapezoid. ; true

    (8) A rectangle is a square. ; false

    (9) A square is a rhombus. ; true

    (10) A square is a trapezoid. ; true

#2

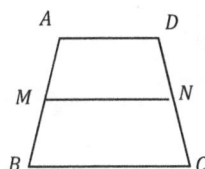

For trapezoid $\square ABCD$, the length of the median $\overline{MN}$ is 10 and the length of $\overline{BC}$ is 2 greater than twice the length of $\overline{AD}$. Find the lengths of the bases.

$\because BC = 2AD + 2.$

$MN = \frac{AD+BC}{2} = \frac{AD+2AD+2}{2} = \frac{3AD+2}{2}$ $\therefore \frac{3AD+2}{2} = 10$; $3AD = 18$; $AD = 6$

$\therefore BC = 2 \cdot 6 + 2 = 14$

Therefore, the lengths of the bases are 6 and 14.

#3 For the following isosceles trapezoids $\square ABCD$, find the measure of $\angle a$.

    (1)

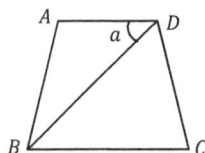

$m \angle C = 75^{\circ},\ \ m \angle ABD = 25^{\circ}$

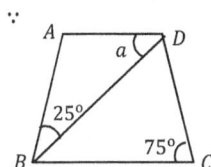

Since $\square ABCD$ is an isosceles trapezoid, $m \angle B = m \angle C = 75^{\circ}$
$\therefore\ m \angle DBC = 75^{\circ} - 25^{\circ} = 50^{\circ}$
Since $\overline{AD} \mathbin{/\!/} \overline{BC}$, $m \angle a = m \angle DBC$ by alternate interior angles .

Therefore, $m \angle a = 50^{\circ}$

(2)

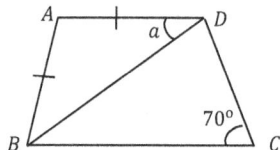

$m \angle C = 70^\circ, \quad \overline{AB} \cong \overline{AD}$

∵

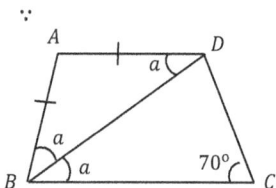

Since $\overline{AB} \cong \overline{AD}$, $m \angle a = m \angle ABD$
Since $\angle a \cong \angle DBC$ by alternate interior angles, $m \angle B = 2\,m \angle a$
Since $\square ABCD$ is an isosceles trapezoid, $\angle B \cong \angle C$
So $m \angle B = 70^\circ$

Therefore, $m \angle a = \dfrac{70^\circ}{2} = 35^\circ$

(3)

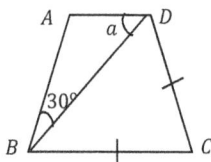

$m \angle ABD = 30^\circ, \quad \overline{BC} \cong \overline{CD}$

∵

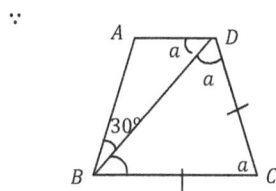

Since $\overline{BC} \cong \overline{CD}$, $\angle DBC \cong \angle BDC$
Since $\overline{AD} /\!/ \overline{BC}$, $\angle a \cong \angle DBC$.
Since $\angle D$ and $\angle B$ are supplementary,

$30^\circ + 3\,m \angle a = 180^\circ$.   ∴ $m \angle a = \dfrac{150^\circ}{3} = 50^\circ$

#4 For the following isosceles trapezoids $\square ABCD$, answer the following :

(1)

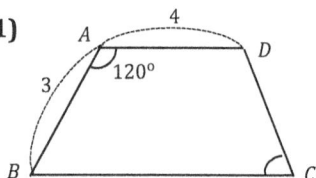

Find the length of the side $\overline{BC}$.

∵

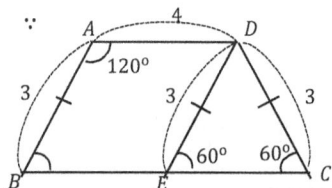

Since $\angle A$ and $\angle C$ are supplementary, $m \angle C = 60^\circ$.
Note that $\overline{DE} \cong \overline{DC}$. Then $\triangle DEC$ is an isosceles triangle.
∴ $m \angle DEC = m \angle C = 60^\circ$

∴ $m \angle EDC = 180^\circ - (60^\circ + 60^\circ) = 60^\circ$; $\overline{EC} \cong \overline{ED} \cong \overline{CD}$
Therefore, $BC = 4 + 3 = 7$.

(2)

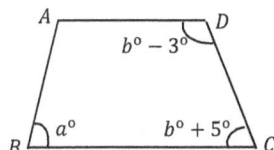

Find the values of a and b.

$\because$ Since $\square ABCD$ is an isosceles trapezoid, $\angle B \cong \angle C$.

So, $a° = b° + 5°$

Since $\angle B$ and $\angle D$ are supplementary, $a° + (b° - 3°) = 180°$

$\therefore (b° + 5°) + (b° - 3°) = 180°$; $2b° = 178°$; $b° = 89°$

Therefore, $a° = b° + 5° = 89° + 5° = 94°$; Hence $a = 94$ and $b = 89$

#5 For the following parallelograms $\square ABCD$, answer the following :

(1)

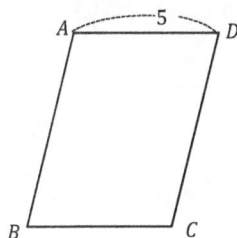

The perimeter of $\square ABCD$ is 30.

The length of $\overline{AD}$ is 5. Find the length of $\overline{AB}$.

$\because$ Since $\overline{AD} \cong \overline{BC}$ and $\overline{AB} \cong \overline{DC}$, $30 = 10 + 2 \cdot AB$; $20 = 2AB$; $AB = 10$

$\therefore$ The length of $\overline{AB}$ is 10.

(2)

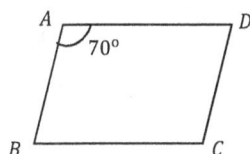

$m \angle A = 70°$.

Find the measures of $\angle C$ and $\angle D$.

$\because$

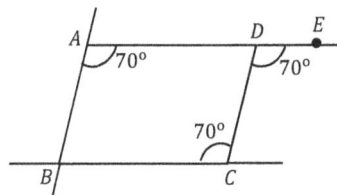

$\because$ Since the opposite angles of a parallelogram are congruent,

$m \angle C = 70°$

Since $\angle DAB \cong \angle EDC$ by corresponding angles,

$m \angle D = m \angle ADC = 180° - 70° = 110°$

(3)

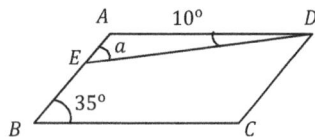

$m \angle B = 35^\circ, \; m \angle ADE = 10^\circ.$

Find $m \angle a$.

∵

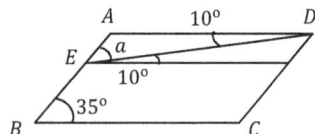

Since $\angle B \cong \angle D$, $m \angle EDC = 35^\circ - 10^\circ = 25^\circ$.

Since $\angle EDC$ and $\angle a$ are corresponding angles,

$m \angle a = 25^\circ$.

(4)

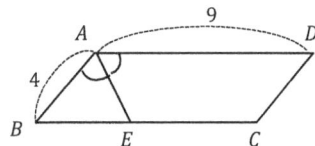

$\angle BAE \cong \angle DAE$

The lengths of $\overline{AD}$ and $\overline{AB}$ are 9 and 4, respectively.
Find the length of $\overline{EC}$.

∵ By alternate interior angles, $m \angle DAE = \mathrm{m} \angle AEB$.

Since $\angle BAE \cong \angle DAE$, $\angle BAE \cong \angle BEA$

So $\triangle BAE$ is an isosceles triangle. ∴ $BE = AB = 4$

Since $\square ABCD$ is a parallelogram, $\overline{AD} \cong \overline{BC}$

Therefore, $EC = BC - BE = 9 - 4 = 5$

(5)

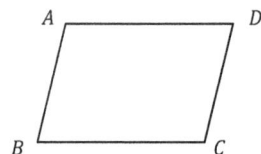

$\angle A : \angle B = 3 : 2$

Find the measure of $\angle C$.

∵ Since the two consecutive angles of a parallelogram are supplementary, $m \angle A + m \angle B = 180^\circ$

Since $\angle A : \angle B = 3 : 2$, $m \angle A = \frac{3}{5} \times 180^\circ = 108^\circ$

Since any two opposite angles of a parallelogram are congruent, $\angle A \cong \angle C$

Therefore, $m \angle C = m \angle A = 108^\circ$

(6)

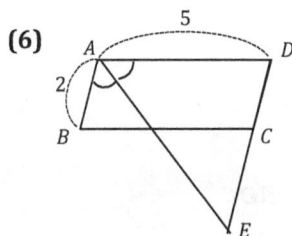

$\angle BAE \cong \angle EAD$

Find the length of $\overline{CE}$.

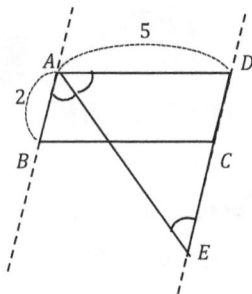

Since $\overline{AB} /\!/ \overline{DE}$,

$\angle BAE \cong \angle AED$ by alternate interior angles.

$\therefore \triangle DAE$ is isosceles. $\therefore \overline{AD} \cong \overline{DE}$

$\therefore DE = DC + CE = 5$

Since $\overline{AB} \cong \overline{CD}$, $CD = 2$. Therefore, $CE = 5 - 2 = 3$

#6 For the parallelogram $\square ABCD$, answer the following :

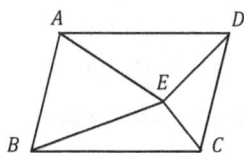

(1) The area of $\triangle ABE$ is 15 $in.^2$ The area of $\triangle BCE$ is 10 $in.^2$
The area of $\triangle CDE$ is 8 $in.^2$ Find the area of $\triangle ADE$.

$15 + 8 = 10 + $ Area of $\triangle ADE$ $\therefore$ Area of $\triangle ADE = 13\ in.^2$

(2) The area of $\square ABCD$ is 50 $in.^2$ Find the sum of the area of $\triangle ABE$ and the area of $\triangle CDE$.

The sum is $\frac{50}{2} = 25\ in.^2$

(3) The area of $\square ABCD$ is 80 $in.^2$ The area of $\triangle ABE$ is 30 $in.^2$ Find the area of $\triangle CDE$.

The sum of the area of $\triangle ABE$ and the area of $\triangle CDE$ is $\frac{1}{2} \times 80 = 40$.

$\therefore$ Area of $\triangle CDE$ is $40 - 30 = 10\ in.^2$

#7 For the following rhombi $\square ABCD$**, answer the following :**

(1)

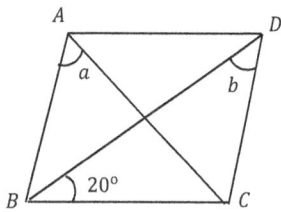

$m \angle DBC = 20^o$

Find the measures of $\angle BAD$ **and** $\angle BDC$**.**

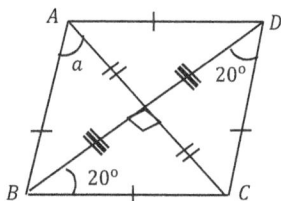

Since $\square ABCD$ is a rhombus , $\overline{AB} \cong \overline{BC} \cong \overline{CD} \cong \overline{AD}$.

∴ $\triangle BDC$ is isosceles. So, $m \angle BDC = 20^o$

∴ $m \angle ABD = m \angle ADB = m \angle DBC = m \angle BDC = 20^o$

Therefore , $m \angle BAD = m \angle A = 180^o - (20^o + 20^o) = 140^o$

(2)

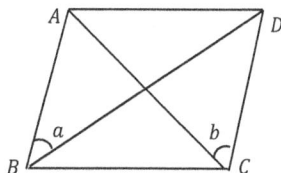

Find the sum of $m \angle a + m \angle b$**.**

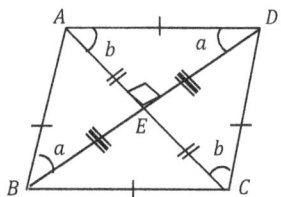

Since the diagonals are perpendicular to each other

and bisect each other, $\angle a \cong \angle ADB$ and $\angle b \cong \angle DAC$.

Since $\triangle ADE$ is a right triangle, $m \angle a + m \angle b = 180^o - 90^o = 90^o$.

(3)

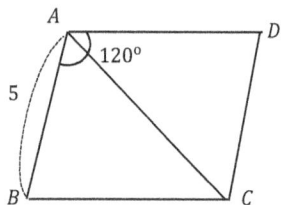

The length of $\overline{AB}$ **is 5 and**

$m \angle A = 120^o$**.**

Find the length of $\overline{AC}$ **.**

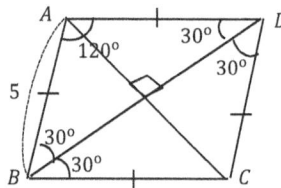

Since $\overline{AB} \cong \overline{AD}$, $\triangle ABD$ is an isosceles triangle .

So $\angle ABD \cong \angle ADB$

∴ $m \angle ABD = \frac{180^o - 120^o}{2} = 30^o$

Since the diagonals are perpendicular and bisect each other,

$m \angle B = m \angle BAC = m \angle BCA = 60^o$ ∴ $\triangle ABC$ is a regular triangle. So, $\overline{AB} \cong \overline{BC} \cong \overline{AC}$.

Therefore, the length of $\overline{AC}$ is 5.

#8 For the following rectangles $\square ABCD$, answer the following :

(1)

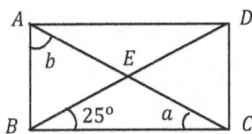

Find $m\angle a$ and $m\angle b$.

Since $\square ABCD$ is a rectangle, $\overline{AE} \cong \overline{BE} \cong \overline{CE} \cong \overline{DE}$.

Since $\overline{BE} \cong \overline{CE}$, $\triangle BCE$ is an isosceles triangle. $\therefore m\angle a = 25°$

Since $\triangle ABC$ is a right triangle , $m\angle b + m\angle a = 90°$

$\therefore m\angle b = 90° - 25° = 65°$

(2)

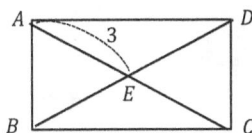

The length of $\overline{AE}$ is 3 inches.

Find the length of $\overline{BD}$.

Since $\overline{AE} \cong \overline{BE} \cong \overline{CE} \cong \overline{DE}$, the length of $\overline{BD}$ is $3 + 3 = 6$ inches.

(3)

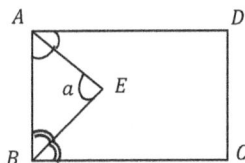

$\angle BAE \cong \angle DAE$ and $\angle ABE \cong \angle EBC$.

Find $m\angle a$.

Since $m\angle A = m\angle B = 90°$, $m\angle EAB = 45°$ and $m\angle ABE = 45°$

Therefore, $m\angle a = 180° - (45° + 45°) = 90°$

#9 For the following squares $\square ABCD$, answer the following :

(1)

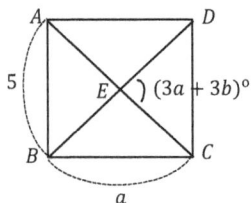

The length of $\overline{AB}$ is 5.

For the length a of $\overline{BC}$,

find the value of b when $m \angle DEC = (3a + 3b)^\circ$

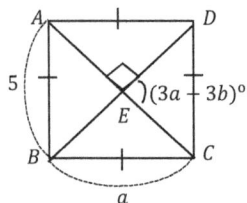

Since $\overline{AB} \cong \overline{BC}$, the length of $\overline{BC}$ is $a = 5$.

Since $(3a + 3b)^\circ = 90^\circ$, $3a + 3b = 90$. So, $3b = 90 - 15 = 75$.

Therefore, $b = 25$.

(2)

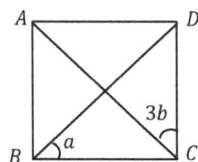

Find the measures of $\angle a$ and $\angle b$.

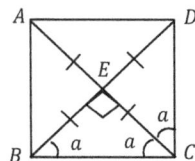

Since $\overline{EB} \cong \overline{EC} \cong \overline{ED} \cong \overline{EA}$, $m \angle EBC = m \angle ECB$.

Since $m \angle BED = 90^\circ$, $m \angle EBC + m \angle ECB = 90^\circ$.

So, $m \angle a = m \angle EBC = 45^\circ$

Since $m \angle 3b = m \angle a$, $m \angle b = \dfrac{45^\circ}{3} = 15^\circ$

(3)

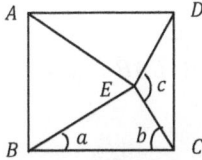

$\overline{AB} \cong \overline{AE} \cong \overline{BE}.$

Find the measures of $\angle a$, $\angle b$, **and** $\angle c$.

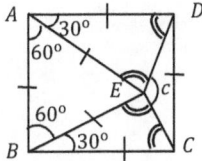

Since $\overline{AB} \cong \overline{AE} \cong \overline{BE}$, $\triangle ABE$ is a regular triangle.

$\therefore m \angle EAB = m \angle ABE = m \angle AEB = 60^\circ$

Since $m \angle A = m \angle B = 90^\circ$, $m \angle a = 30^\circ$ and $m \angle EAD = 30^\circ$.

Since $\triangle BEC$ and $\triangle AED$ are isosceles triangles,

$\angle b \cong \angle BEC \cong \angle AED \cong \angle ADE$

So, $m \angle b \cong m \angle BEC \cong m \angle AED \cong m \angle ADE = \frac{180^\circ - 30^\circ}{2} = 75^\circ$

Since $m \angle C = m \angle D = 90^\circ$, $m \angle ECD = m \angle EDC = 90^\circ - 75^\circ = 15^\circ$

Therefore, $m \angle c = 180^\circ - (15^\circ + 15^\circ) = 150^\circ$

#10 Determine whether the following statements are true or false.

(1) **A rhombus with congruent diagonals is a square.** ; true

(2) **A parallelogram with one right angle is a rectangle.** ; true

(3) **The diagonals of a rectangle are perpendicular to each other.** ; false

(4) **A parallelogram is a rectangle.** ; false

(5) **A parallelogram with diagonals perpendicular to each other is a rhombus.** ; true

Solutions for Chapter 4

#1 Given a circle with center P, let $\overline{CB} = \overline{PB}$. Find $m\,\widehat{CB}\,$; $m\,\widehat{AC}\,$; $m\,\widehat{ACB}\,$; $m\,\widehat{CAB}\,$.

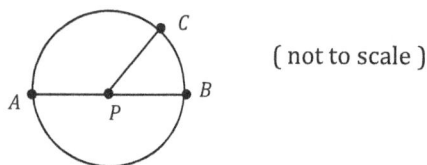

(not to scale)

($\because$

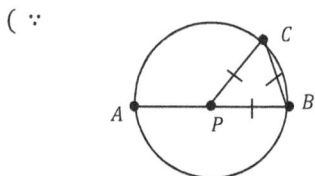

Since $\overline{CB} = \overline{PB}$ and $\overline{PC} = \overline{PB}$ (radii) , $\triangle\,PBC$ is a regular triangle.

$\therefore\; m\,\angle CPB = \dfrac{180^{\circ}}{3} = 60^{\circ}\;$ $\therefore\; m\,\widehat{CB} = 60^{\circ}$

Since $m\,\angle CPA = 180^{\circ} - m\,\angle CPB = 120^{\circ}$, $m\,\widehat{AC} = 120^{\circ}$.

So $m\,\widehat{ACB} = m\,\widehat{AC} + m\,\widehat{CB} = 120^{\circ} + 60^{\circ} = 180^{\circ}$

(OR simply $m\,\widehat{ACB} = 180^{\circ}$ because $\widehat{ACB}$ is a semicircle)

Also, $m\,\widehat{CAB} = 360^{\circ} - 60^{\circ} = 300^{\circ}$.   )

#2 Given a circle with center P, let $\overline{AC} = \overline{BD}$ be diameters of the circle and $m\,\angle DBC = 30^{\circ}$. Find the measure of each minor arc of the circle.

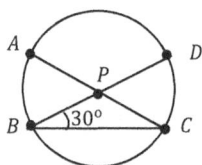

Since $\overline{PB}$ and $\overline{PC}$ are radii of the circle, $\triangle\,PBC$ is isosceles.

So $m\,\angle PBC = m\,\angle PCB = 30^{\circ}$

$\therefore\; m\,\angle BPC = 180^{\circ} - (30^{\circ} + 30^{\circ}) = 120^{\circ}$

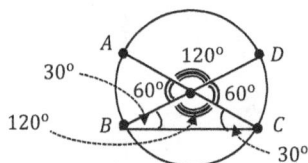

Thus , using vertical angles, and supplement angles,

$m\,\widehat{AB} = 60^{\circ}, m\,\widehat{BC} = 120^{\circ}, m\,\widehat{CD} = 60^{\circ}$,

and $m\,\widehat{AD} = 120^{\circ}$

#3 The point *P* is the circumcenter of the triangles △ *ABC* . Find the measure of ∠*a* for each triangle.

(1)

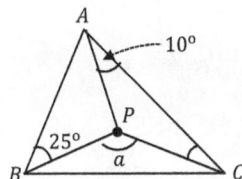

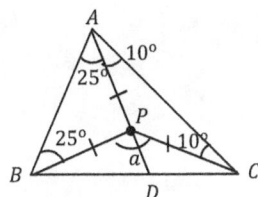

$m \angle a = m \angle BPD + m \angle CPD$

$= (25° + 25°) + (10° + 10°) = 50° + 20° = 70°$

(2)

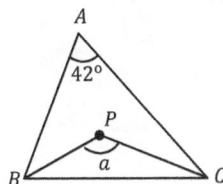

$m \angle a = 2 \, m \angle A$

$= 2 \times 42° = 84°$

(3)

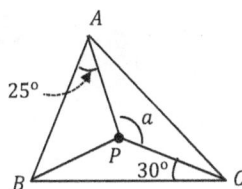

$m \angle a = 2 \, m \angle B$

$= 2 \times (25° + 30°) = 110°$

(4)

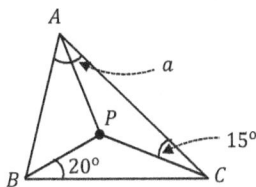

$m \angle B = \dfrac{1}{2} (150°) = 75°$

$\therefore \; m \angle PAB = m \angle PBA = 55°$

$\therefore \; m \angle a = 55° + 15° = 70°$

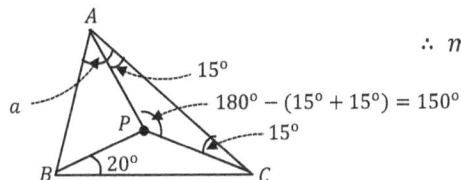

$180° - (15° + 15°) = 150°$

(5)

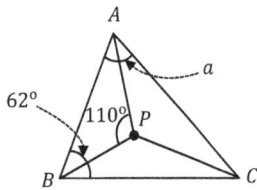

Since $m \angle C = \frac{1}{2} m \angle APB = 55^\circ$

From a triangle $\triangle ABC$,

$\therefore m \angle a = 180^\circ - (62^\circ + 55^\circ) = 63^\circ$

(6)

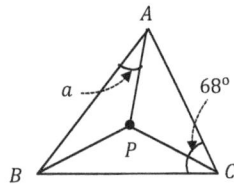

$2 \, m \angle a = 180^\circ - 136^\circ = 44^\circ$

$\therefore m \angle a = 22^\circ$

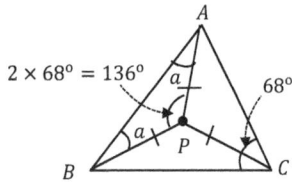

#4 The point Q is the incenter of the triangles ABC. Find the measure of $\angle a$ for each triangle.

(1)

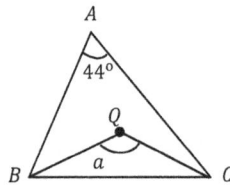

$m \angle a = 90^\circ + \frac{1}{2} m \angle A = 90^\circ + 22^\circ = 112^\circ$

(2)

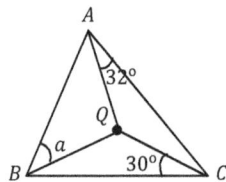

$m \angle a + 30^\circ + 32^\circ = 90^\circ$

$\therefore m \angle a = 28^\circ$

(3)

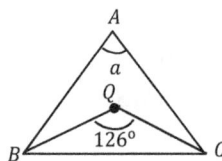

$126^\circ = 90^\circ + \frac{1}{2} m \angle a$

$\therefore \frac{1}{2} m \angle a = 36^\circ \quad \therefore m \angle a = 72^\circ$

(4)

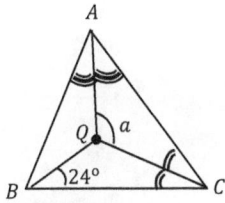

$m \angle QAB + 24^\circ + m \angle QCA = 90^\circ \therefore m \angle QAB + m \angle QCA = 66^\circ$

From $\triangle QAC$, $m \angle a + m \angle QAC + m \angle QCA = 180^\circ$

Since $m \angle QAB = m \angle QAC$, $m \angle a + m \angle QAB + m \angle QCA = 180^\circ$

$\therefore m \angle a + 66^\circ = 180^\circ$; $m \angle a = 114^\circ$

(5)

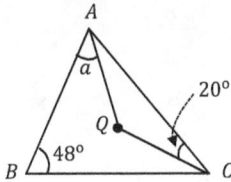

$m \angle QBC + 20^\circ + m \angle a = 90^\circ$

Since Q is incenter of $\triangle ABC$, $m \angle QBC = \frac{1}{2}(48^\circ) = 24^\circ$

$\therefore m \angle a = 90^\circ - (24^\circ + 20^\circ) = 46^\circ$

#5 For the following problems, the point Q is the incenter of the triangle $\triangle ABC$. Let $\overline{DE} // \overline{BC}$.

(1)

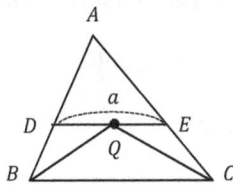

Let $AD = 4$; $DB = 2$; $AE = 5$; $EC = 3$.

Find the value of a and the perimeter of $\triangle ADE$.

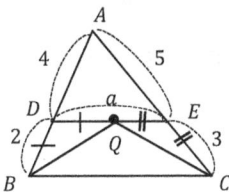

$(\because a = DQ + QE$

Since $\overline{DQ} \cong \overline{DB}$ and $\overline{QE} \cong \overline{EC}$, $a = 2 + 3 = 5$.

$\therefore$ The perimeter of $\triangle ADE$ is $5 + 4 + 5 = 14$)

(2)

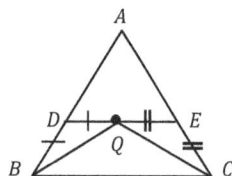

Let $AB = 10$; $AC = 8$; $BC = 6$.

Find the perimeter of $\triangle ADE$.

($\because$ Since $\overline{DQ} \cong \overline{DB}$ and $\overline{QE} \cong \overline{EC}$, the perimeter of $\triangle ADE$ is

$AD + DE + AE = AD + (DQ + QE) + AE = AD + (DB + EC) + AE$

$= (AD + DB) + (EC + AE) = AB + AC = 10 + 8 = 18$)

#6

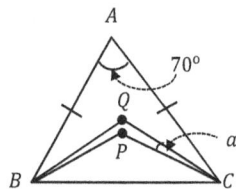

The point P and Q are the circumcenter and incenter of the triangle $\triangle ABC$, respectively.

Let $\overline{AB} \cong \overline{AC}$ and $m \angle A = 70°$.

Find the measure of $\angle a$.

($\because$ Since P is the circumcenter of $\triangle ABC$, $m \angle BPC = 2m \angle A = 140°$ and $\overline{BP} \cong \overline{PC}$.

So, $\angle PBC \cong \angle PCB$ $\quad \therefore$ $\angle PCB = \frac{180° - 140°}{2} = 20°$

Since Q is the incenter of $\triangle ABC$, $m \angle BQC = 90° + \frac{1}{2}m \angle A = 90° + 35° = 125°$ and $\angle QBC \cong \angle QCB$.

$\therefore$ $\angle QCB = \frac{180° - 125°}{2} = 27.5°$ $\quad$ Therefore, $m \angle a = 27.5° - 20° = 7.5°$)

#7

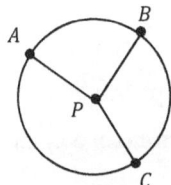

For a circle with center P,

$$m\,\widehat{AB} : m\,\widehat{BC} : m\,\widehat{CA} = 2 : 3 : 4.$$

Find the measures of $\angle APB : \angle BPC : \angle APC$.

($\because$ Since $m\,\angle APB = m\,\widehat{AB}, m\,\angle BPC = m\,\widehat{BC}$, and $m\,\angle APC = m\,\widehat{AC}$,

$m\,\angle APB : m\,\angle BPC : m\,\angle APC = 2 : 3 : 4$

$\therefore\ m\,\angle APB = 360^\circ \times \frac{2}{9} = 80^\circ\ \ m\,\angle BPC = 360^\circ \times \frac{3}{9} = 120^\circ\ \ $ and $m\,\angle APC = 360^\circ \times \frac{4}{9} = 160^\circ$)

#8

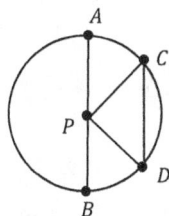

For a circle with center P, $\overline{AB} \parallel \overline{CD}$ and

$$m\,\widehat{BD} = \tfrac{1}{2} m\,\widehat{CD}$$

Find the measure of $\angle CPD$.

($\because$ Since $m\,\widehat{BD} = \frac{1}{2} m\,\widehat{CD}$, $m\,\angle BPD = \frac{1}{2} m\,\angle CPD$; $m\,\angle CPD = 2\,m\,\angle BPD$

Let $m\,\angle BPD = a^\circ$. Then $m\,\angle CPD = 2\,a^\circ$

Since $\overline{AB} \parallel \overline{CD}$, $\angle BPD \cong \angle PDC$ (alternate interior angles) So, $m\,\angle BPD = m\,\angle PDC = a^\circ$

Since $\overline{PC} = \overline{PD}$ (; radii), $\triangle PCD$ is isosceles. So, $\angle PDC \cong \angle PCD$ Thus, $m\,\angle PDC = m\,\angle PCD = a^\circ$

Since $m\,\angle CPD + m\,\angle PCD + m\,\angle PDC = 180^\circ$, $2\,a^\circ + a^\circ + a^\circ = 180^\circ$

Thus, $4a^\circ = 180^\circ$; $a^\circ = 45^\circ$

Therefore, $m\,\angle CPD = 2\,a^\circ = 90^\circ$)

#9

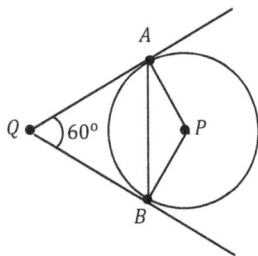

Two tangents through an exterior point Q intersect a circle (with center P) at points A and B. The measure of $\angle Q$ is 60°. Find the measure of $\angle PBA$.

($\because$ Since $\overline{QA} \cong \overline{QB}$, $\triangle QAB$ is isosceles. $\therefore$ $m \angle QAB = m \angle QBA$

Since $m \angle Q = 60°$, $m \angle ABQ = 60°$

Since $\angle ABQ$ is an inscribed angle intercepting the arc $\widehat{AB}$, $m \angle ABQ = \frac{1}{2} m \widehat{AB}$

$\therefore$ $m \widehat{AB} = 2 \times 60° = 120°$

Since $m \angle APB = m \widehat{AB}$, $m \angle APB = 120°$

Since $\triangle APB$ is isosceles, $m \angle PAB = m \angle PBA$ Therefore, $m \angle PBA = 30°$)

#10

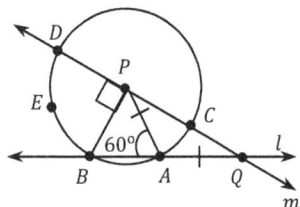

Secants l and m, through an exterior point Q, intersect a circle (with center P) at points A and B ; C and D, respectively, where $\overline{CD}$ is a diameter of the circle. Let $m \angle PAB = 60°$, $\overline{QA} \cong \overline{AP}$, and $m \widehat{DEB} = 3$ inches. Find $m \widehat{AC}$.

($\because$ Since $\angle PAB$ is an exterior angle of $\triangle APQ$, $m \angle PAB = m \angle APQ + m \angle AQP$

Since $\overline{QA} \cong \overline{AP}$, $\triangle PAQ$ is isosceles. $\therefore$ $m \angle APQ = m \angle AQP$ Thus, $m \angle APQ = m \angle AQP = 30°$

Since $\angle BPD$ is a central angle, $m \angle BPD = m \widehat{BED} = 3$ inches.

Since $m \angle BPD = 90°$, $90° = 3$ inches.

Since $\angle APC$ is a central angle, $m \angle APC = m \widehat{AC}$

So, $m \widehat{AC} = m \angle APC = m \angle APQ = 30° = 1$ inch.

Therefore, $m \widehat{AC} = 1$ inch.)

#11

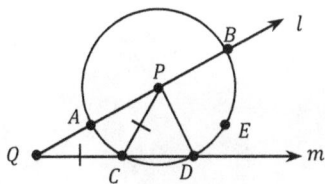

Secants l and m, through an exterior point Q, intersect a circle (with center P) at points A and B ; C and D , respectively, where $\overline{AB}$ is a diameter of the circle.

Let $\overline{QC} \cong \overline{CP}$ and $m\,\widehat{AC} = a^\circ$.

Find $m\,\widehat{BED}$.

($\because$ Since $\overline{QC} \cong \overline{CP}$, $\triangle\,QCP$ is isosceles. ; $m\,\angle PQC = m\,\angle QPC$

Since $\angle APC$ is a central angle, $m\,\angle APC = m\,\widehat{AC} = a^\circ$

$m\,\angle PQC = m\,\angle QPC = m\,\angle APC = a^\circ$

Since $\angle PCD$ is an exterior angle of $\triangle\,PQC$, $m\,\angle PCD = a^\circ + a^\circ = 2a^\circ$

Since $\overline{CP} \cong \overline{DP}$ (radii), $\triangle\,PCD$ is isosceles.

$\therefore$ $m\,\angle PCD = m\,\angle PDC = 2a^\circ$

Since $\angle BPD$ is an exterior point of $\triangle\,PQD$, $m\,\angle BPD = \angle PQD + \angle PDQ = a^\circ + 2a^\circ = 3a^\circ$

Since $\angle BPD$ is a central angle, $m\,\widehat{BED} = m\,\angle BPD$

Therefore, $m\,\widehat{BED} = 3a^\circ$)

#12

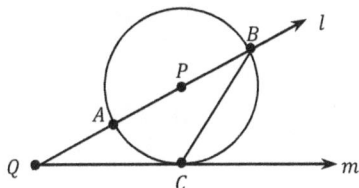

A secant l and a tangent m, through an exterior point Q, intersect a circle (with center P) at points A and B ; C, respectively, where $\overline{AB}$ is a diameter of the circle.

(1) Let $m \angle Q = 32°$. Find $m \angle QBC$.

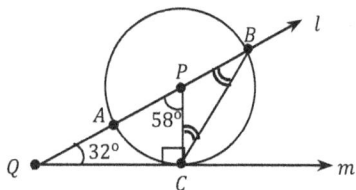

(∵ Since m is a tangent of the circle, $\overline{PC} \perp \overline{QC}$ ∴ $m \angle PCQ = 90°$

Since $m \angle Q = 32°$, $m \angle QPC = 180° - (90° + 32°) = 58°$

Since $\angle QPC$ is an exterior angle of $\triangle PCB$, $m \angle QPC = m \angle PCB + m \angle PBC$.

Since $\overline{PC} \cong \overline{PB}$ (radii), $m \angle PCB = m \angle PBC$.

So, $m \angle PBC = m \angle PCB = 29°$

Therefore, $m \angle QBC = m \angle PBC = 29°$.)

(2) Let $m \angle APC = 70°$. Find $m \angle ACQ$.

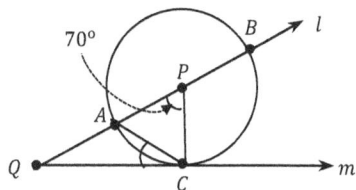

(∵ Since $\angle APC$ is a central angle, $m \angle APC = m \widehat{AC}$ ∴ $m \widehat{AC} = 70°$

Since $\angle ACQ$ is an inscribed angle, $m \angle ACQ = \frac{1}{2} m \widehat{AC} = 35°$.)

(3) Let $m \angle QBC = 35°$. Find $m \angle ACQ$.

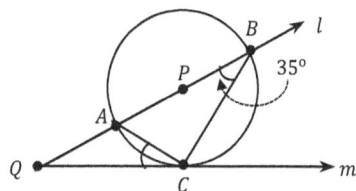

(∵ Since $\angle ABC$ is an inscribed angle, $m \angle ABC = \frac{1}{2} m \widehat{AC}$ ∴ $m \widehat{AC} = 70°$

Therefore, $m \angle ACQ = \frac{1}{2} m \widehat{AC} = 35°$.)

#13 For the following, find the value of :

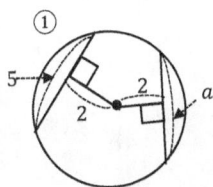

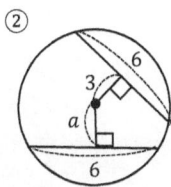

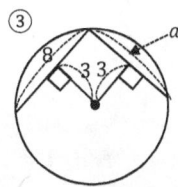

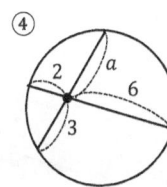

$a = 5$

$a = 3$

$a = 8$

$a = 4$
$(\because 2 \cdot 6 = 3 \cdot a$
$; \; 12 = 3a$
$; \; a = 4 \;)$

#14 For the following, find the measure of $\angle Q$:

(1)

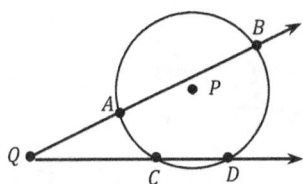

$m \, \widehat{AC} = 2$ and $m \, \widehat{BD} = 6$

$(\because m \angle Q = \dfrac{1}{2} \, (6 - 2) = \dfrac{4}{2} = 2 \;)$

(2)

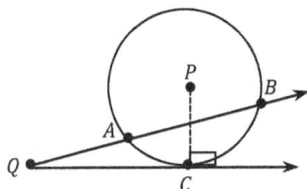

$m \, \widehat{AC} = 3$ and $m \, \widehat{BC} = 5$

$(\because m \angle Q = \dfrac{1}{2} \, (5 - 3) = \dfrac{2}{2} = 1 \;)$

(3)

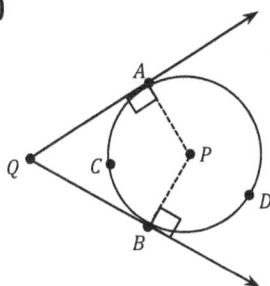

$m \, \widehat{ACB} = 4$ and $m \, \widehat{ADB} = 7$

$(\because m \angle Q = \dfrac{1}{2} \, (7 - 4) = \dfrac{3}{2} \;)$

#15 For the following, find the value of x :

(1)

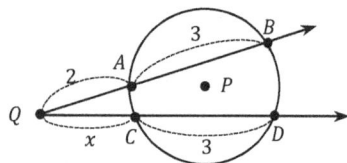

$QA = 2$ and $AB = CD = 3$

($\because$ $QA \cdot QB = QC \cdot QD$

Let $QC = x$. Then, $2 \cdot 5 = x \cdot (x+3)$; $x^2 + 3x - 10 = 0$; $(x-2)(x+5) = 0$

$\therefore x = 2$ or -5

Since $QC > 0$, $x = QC = 2$)

(2)

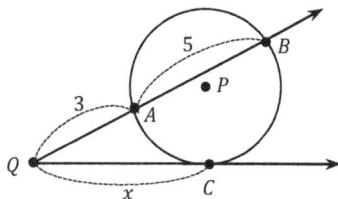

$QA = 3$ and $AB = 5$

($\because$ $QC^2 = 3 \cdot 8 = 24$ $\therefore x = QC = \sqrt{24} = 2\sqrt{6}$)

(3)

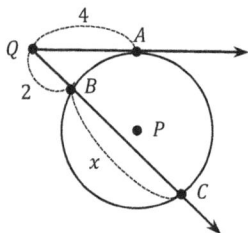

$QA = 4$ and $QB = 2$

($\because$ $4^2 = 2 \cdot (2+x)$; $16 = 4 + 2x$; $12 = 2x$ $\therefore x = 6$)

(4)

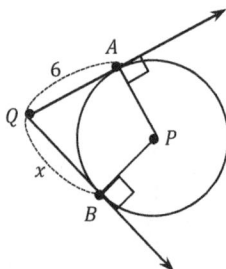

$QA = 6$

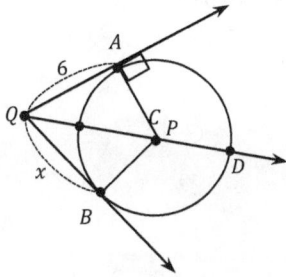

$(\because$ Since $QA^2 = QC \cdot QD$ and $QB^2 = QC \cdot QD$,

$QA^2 = QB^2$; $QA = QB$ $\therefore x = QB = 6$

#16

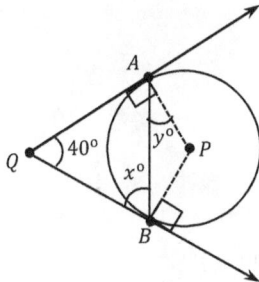

$\overline{PA} \perp \overline{AQ}$, $\overline{PB} \perp \overline{BQ}$.

Let $m \angle Q = 40°$. Find the values of x and y.

$(\because$ Since $\overleftrightarrow{QA}$ and $\overleftrightarrow{QB}$ are tangents, $\overline{QA} \cong \overline{QB}$ $\therefore \triangle AQB$ is isosceles.

So, $m \angle QBA = m \angle QAB$ $\therefore 40° + x° + x° = 180°$; $2x° = 140°$; $x° = 70°$ $\therefore x = 70$

Also, $m \angle APB = m \widehat{AB}$ and $x° = m \angle ABQ = \frac{1}{2} m \widehat{AB}$; $70° = \frac{1}{2} m \widehat{AB}$; $m \widehat{AB} = 140°$

Since $\triangle PAB$ is isosceles, $y° = 20°$ $\therefore y = 20$

#17 For the following circles with centers P, find the measure of $\angle a$:

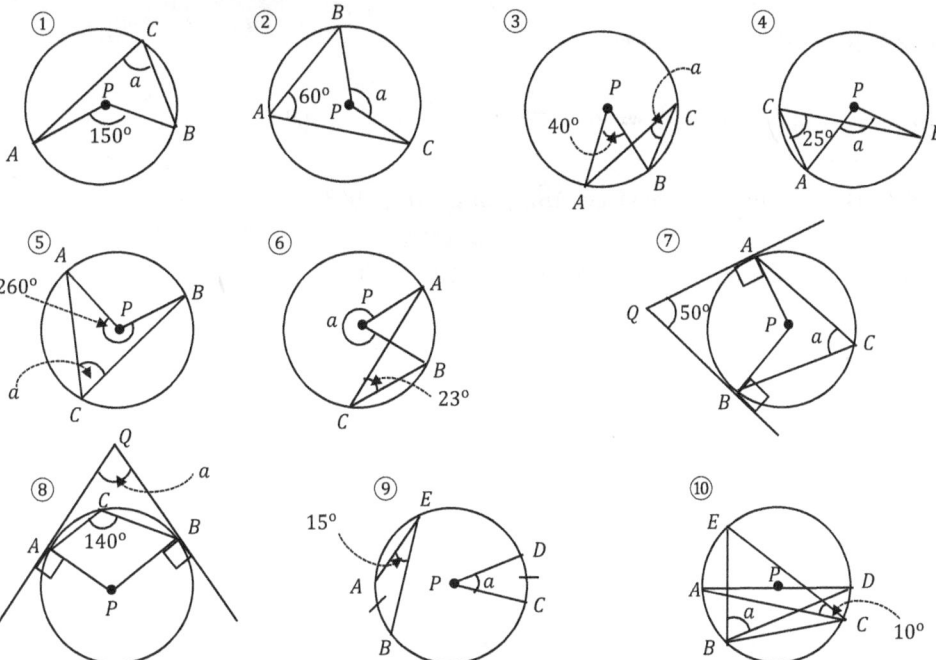

① $m \angle a = \frac{1}{2} m \, \overparen{AB}$ and $m \angle APB = \frac{1}{2} m \, \overparen{AB}$ $\therefore m \angle a = \frac{1}{2} \cdot 150° = 75°$

② $m \angle a = 2 \cdot 60° = 120°$

③ $m \angle a = \frac{1}{2} \cdot 40° = 20°$

④ $m \angle a = 2 \cdot 25° = 50°$

⑤ $m \angle APB = 360° - 260° = 100°$ $\therefore m \angle a = \frac{1}{2} \cdot 100° = 50°$

⑥ $m \angle APB = 2 \cdot 23° = 46°$ $\therefore m \angle a = 360° - 46° = 314°$

⑦ $m \angle a = \frac{1}{2} \cdot m \angle APB$

   Since $m \angle APB + m \angle Q = 180°$, $m \angle APB = 130°$ $\therefore m \angle a = 65°$

⑧

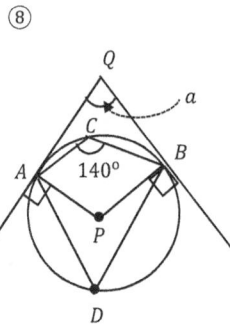

Since $m \angle ACB + m \angle ADB = 180°$,

$m \angle ADB = 40°$

$\therefore m \, \overparen{AB} = 80°$ $\therefore m \angle APB = 80°$

Since $m \angle a + m \angle APB = 180°$, $m \angle a = 100°$

⑨ Since $m \angle AEB = 15°$, $m \, \overparen{AB} = 30°$

   Since $\overparen{AB} \cong \overparen{CD}$, $m \, \overparen{CD} \cong m \, \overparen{AB} = 30°$ Therefore, $m \angle a = 30°$

⑩ Since $\angle ABD$ is inscribed in a semicircle $\overparen{ABD}$, $m \angle ABD = 90°$

   Since $m \angle ABE = m \angle ACE = 10°$, $m \angle a = 90° - 10° = 80°$

#18 **For the following circles (P is the center), find the measures of $\angle a$, $\angle b$, and $\angle c$:**

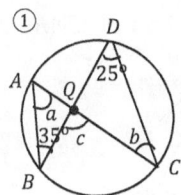

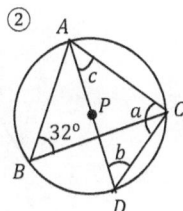

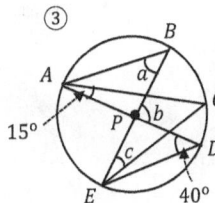

① Since $\angle BAC$ and $\angle BDC$ are inscribed angles intercepting arc $\overset{\frown}{BC}$,

$m\,\angle BAC = m\,\angle BDC \ \therefore\ m\,\angle a = 25°$

Since $\angle ABD$ and $\angle ACD$ are inscribed angles intercepting arc $\overset{\frown}{AD}$,

$m\,\angle ABD = m\,\angle ACD \ \therefore\ m\,\angle b = 35°$

Since $\angle c$ is an exterior angle of $\triangle ABQ$, $m\,\angle c = m\,\angle a + 35° \ \therefore\ m\,\angle c = 25° + 35° = 60°$

② Since $\angle a$ is an inscribed angle in a semicircle $\overset{\frown}{ACD}$, $m\,\angle a = 90°$

Since $\angle B$ and $\angle ADC$ are inscribed angles intercepting arc $\overset{\frown}{AC}$, $m\,\angle b = m\,\angle ADC = 32°$

$\therefore\ m\,\angle c = 180° - (m\,\angle a + m\,\angle b) = 180° - 90° - 32° = 58°$

③ Since $\angle ABE$ and $\angle ADE$ are inscribed angles intercepting arc $\overset{\frown}{AE}$, $m\,\angle a = 40°$

Since $\triangle APB$ is isosceles, $m\,\angle b = m\,\angle a + m\,\angle BAP = 40° + 40° = 80°$

Since $\angle b$ is an exterior angle of $\triangle APB$,

$m\,\angle b = 15° + m\,\angle BAC + m\,\angle a = 15° + m\,\angle BAC + 40° = 55° + m\,\angle BAC.$

$\therefore m\,\angle BAC = m\,\angle b - 55° = 80° - 55° = 25°$

Since $\angle BAC$ and $\angle C$ are inscribed angles intercepting arc $\overset{\frown}{BC}$,

$m\,\angle c = m\,\angle BAC = 25°$

#19 For the following circumscribed quadrilaterals $\square ABCD$ about a circle, find the value of x :

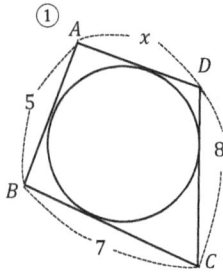

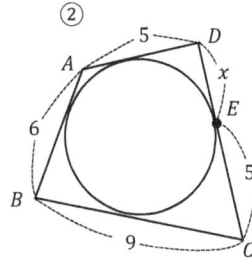

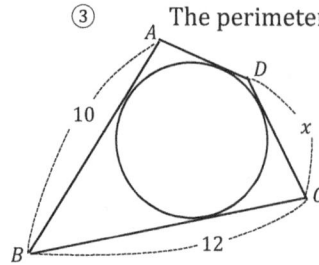

① $AB + DC = BC + AD$; $5 + 8 = 7 + x$; $x = 13 - 7 = 6$

② Since $AB + DC = BC + AD$, $DC = 8$

     Since $CE = 5$, $DE = 8 - 5 = 3$; $x = 3$

③ Since $10 + DC = 12 + AD$,

     the perimeter of $\square ABCD$ is $30 = AB + BC + CD + DA = (AB + CD) + (BC + DA)$

$$= (BC + AD) + (BC + AD) = 2(BC + AD)$$

$\therefore BC + AD = 15$; $12 + AD = 15$; $AD = 3$

Since $BC + AD = 15$, $AB + DC = 15$ $\therefore DC = 5$; $x = 5$

#20 For the following inscribed quadrilaterals $\square ABCD$ in a circle with center P,

     find the measures of $\angle a$ and $\angle b$ for ① and ③ ; find the measure of $\angle a$ for ② :

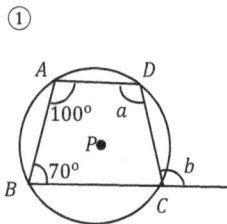

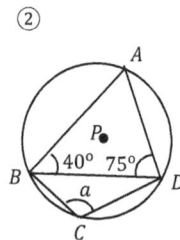

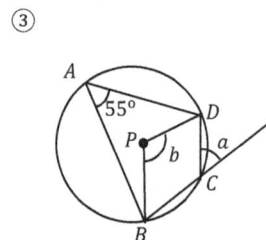

① $m \angle a + 70^\circ = 180^\circ$; $m \angle a = 110^\circ$ and $m \angle b = m \angle A = 100^\circ$

② For a triangle $\triangle ABD$, $m \angle A = 180^\circ - (40^\circ + 75^\circ) = 65^\circ$

Since $m \angle A + m \angle a = 180^\circ$, $m \angle a = 180^\circ - m \angle A = 180^\circ - 65^\circ = 115^\circ$

③ $m \angle a = m \angle A = 55^\circ$

Since $m \angle A = \frac{1}{2} m \, \overbrace{BCD}$ and $m \angle b = m \, \overbrace{BCD}$, $m \angle b = 2 \, m \angle A = 110^\circ$

Solutions for Chapter 5

#1

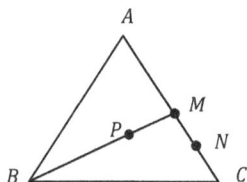

For a triangle $\triangle ABC$, $\overline{BM}$ is a median of $\triangle ABC$ and the point P is the centroid of $\triangle ABC$. Answer the following.

(1) $PM = 2$. Find the length of $\overline{BM}$.

∵ $BP : PM = 2 : 1$ ∴ $BM : PM = 3 : 1 = 6 : 2$ Therefore, $BM = 6$.

(2) $BP = 8$. Find the length of $\overline{PM}$ and $\overline{BM}$.

∵ $BP : PM = 2 : 1 = 8 : 4$ ∴ $PM = 4$ and $BM = 12$.

(3) $A = (2,6)$, $C = (6,4)$. Find the coordinates of M.

∵ Since M is the midpoint of $\overline{AC}$, $M = \left(\frac{2+6}{2}, \frac{6+4}{2}\right) = \left(\frac{8}{2}, \frac{10}{2}\right) = (4, 5)$

(4) A point N is on $\overline{AC}$ and $AN : NC = 3 : 1$. Find the coordinates of N.

∵ $N = \left(\frac{1\cdot2+3\cdot6}{4}, \frac{1\cdot6+3\cdot4}{4}\right) = \left(\frac{20}{4}, \frac{18}{4}\right) = \left(5, \frac{9}{2}\right)$

#2 Given following three lengths of the sides of a triangle, determine whether we can construct a triangle.

(1) $4, 5, 6$ (∵ $6 < 4 + 5$; True. So we can draw a triangle.)

(2) $3, 5, 9$ (∵ $9 \not< 3 + 5$; We cannot draw a triangle.)

(3) $8, 10, 20$ (∵ $20 \not< 8 + 10$; We cannot draw a triangle.)

(4) $2, 2, 2$ (∵ $2 < 2 + 2$; True. So we can draw a triangle.)

(5) $3, 3, 6$ (∵ $6 \not< 3 + 3$; We cannot draw a triangle.)

#3

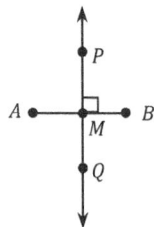

The line $\overleftrightarrow{PQ}$ is the perpendicular bisector of a segment $\overline{AB}$. Determine whether the following expressions are true or false.

(1) $AP = BP$; true

(2) $AB = PQ$; false

(3) $AM = MB$; true

(4) $AP = AQ$; false

(5) $m\angle AMQ = m\angle PMB$; true

(6) $AQ = BQ$; true

#4

For a line l and a point P, the point P is an external point and the point Q is on the line $\overleftrightarrow{PQ}$. $\overleftrightarrow{PQ}$ is perpendicular to a given l, through the point P. Determine whether the following expressions are true or false.

(1) $AP = BP$ **(2)** $AQ = BQ$ **(3)** $AP = AQ$

true   true   false

(4) $AM = BM$ **(5)** $m \angle PMA = m \angle PMB$ **(6)** $PM = MQ$

true   true   false

#5

The point D in the interior of $\angle BAC$ is on the bisector of $\angle BAC$. Determine whether the following expressions are true or false.

(1) $AB = AC$ **(2)** $m \angle BAD = m \angle CAD$ **(3)** $BD = CD$

true   true   true

(4) $AB = BD$ **(5)** $\overline{AB} \perp \overline{BD}$ **(6)** $ED = DF$

false   false   true

Solutions for Chapter 6

#1 **Find the area of the polygons** $\square ABCD$ **.**

(1) The diagonals of rhombus $\square ABCD$ **intersect at point** P **.**

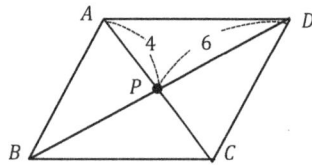

$\therefore$ Area is $\frac{1}{2} \cdot (4+4) \cdot (6+6) = \frac{1}{2} \cdot 8 \cdot 12 = 48$

(2) The trapezoid $\square ABCD$ **has an altitude of 6.**

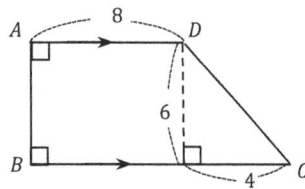

$\therefore$ Area is $\frac{1}{2} h (b_1 + b_2) = \frac{1}{2} \cdot 6 \cdot (8 + 12) = 60$

#2 **A regular polygon has a perimeter of 60 and an apothem of 8. Find the area of the polygon.**

$\therefore$ Area is $A = \frac{1}{2} aP = \frac{1}{2} \cdot 8 \cdot 60 = 240$

#3 **The area of a regular octagon with an apothem of 4 is 160.**

Find the length of each side of the octagon.

$\therefore$ $A = \frac{1}{2} aP = \frac{1}{2} a (n \cdot b) = 160$; $\frac{1}{2} \cdot 4 (8 \cdot b) = 160$; $b = 10$

#4 Find the perimeter and area for the following :

(1)

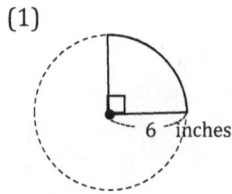

6 inches

(2)

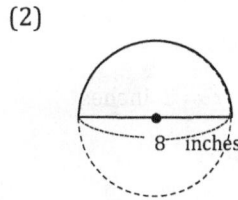

8 inches

$\because$ (1) The circumference of the circle is $C = 2\pi r = 2\pi \cdot 6 = 12\pi$

$\therefore$ Perimeter is $\frac{1}{4} \cdot 12\pi + 6 + 6 = 3\pi + 12$ inches.

The area of the circle is $A = \pi r^2 = \pi 6^2 = 36\pi$ in.^{2}

$\therefore$ The area we are looking for is $\frac{1}{4} \cdot 36\pi = 9\pi$ in.^{2}

(2) $C = 2\pi r = 2\pi \cdot 4 = 8\pi$ $\therefore$ Perimeter is $\frac{1}{2} \cdot 8\pi + 8 = 4\pi + 8$ inches.

$A = \pi r^2 = \pi 4^2 = 16\pi$ in.^{2} $\therefore$ The area we are looking for is $\frac{1}{2} \cdot 16\pi = 8\pi$ in.^{2}

#5 For a circle with a circumference of 14 and the measure of arc $m\,\widehat{AB} = 30$, answer the following.

(1) Find the area of the circle.

$\because$ The circumference is $C = 2\pi r$. $\therefore 14 = 2\pi r$; $r = \frac{14}{2\pi} = \frac{7}{\pi}$

So, the area is $A = \pi r^2 = \pi(\frac{7}{\pi})^2 = \frac{49}{\pi}$

(2) Find the length of $\widehat{AB}$.

$\because L = 2\pi r \cdot \frac{x}{360} = 2\pi r \cdot \frac{30}{360} = 2\pi \cdot \left(\frac{7}{\pi}\right) \cdot \frac{30}{360} = 14 \cdot \frac{1}{12} = \frac{7}{6}$

(3) Find the area of the sector with arc $\widehat{AB}$.

$\because S = \frac{1}{2}rL = \frac{1}{2} \cdot \left(\frac{7}{\pi}\right) \cdot \frac{7}{6} = \frac{49}{12\,\pi}$

OR $S = \pi r^2 \frac{x}{360} = \pi \cdot \left(\frac{7}{\pi}\right)^2 \cdot \frac{30}{360} = \pi \cdot \left(\frac{7}{\pi}\right)^2 \cdot \frac{1}{12} = \frac{49}{12\,\pi}$

#6 Find the length of the arc $\widehat{AB}$ and the area.

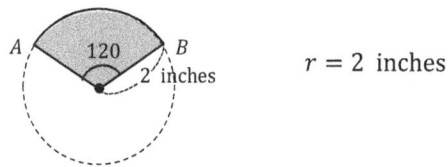

$r = 2$ inches

$$\because \quad L = 2\pi r \cdot \frac{x}{360} = 2\pi(2) \cdot \frac{120}{360} = \frac{4}{3}\pi \quad \therefore m\,\widehat{AB} = \frac{4}{3}\pi \ \text{inches}$$

$$S = \frac{1}{2}rL = \frac{1}{2} \cdot 2 \cdot \frac{4}{3}\pi = \frac{4}{3}\pi \ \text{in.}^2$$

#7 Find the value of x for the following :

(1)   (2)   (3)

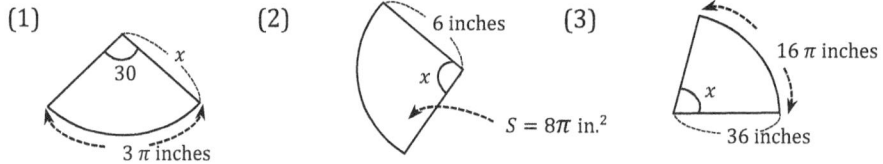

$$\because \ (1) \ L = 2\pi r \cdot \frac{30}{360} \quad ; \quad 3\pi = \frac{\pi r}{6} \quad \therefore \ x = r = 18 \ \text{inches}$$

$$(2) \ S = \pi r^2 \frac{x}{360} \quad ; \quad 8\pi = \pi \cdot 36 \cdot \frac{x}{360} \quad ; \quad 8 = \frac{x}{10} \quad \therefore \ x = 80$$

$$(3) \ S = \frac{1}{2}rL = \frac{1}{2} \cdot 36 \cdot 16\pi = 288\pi$$

$$\therefore \ S = \pi r^2 \frac{x}{360} = 288\pi \quad ; \quad \pi \cdot (36)^2 \cdot \frac{x}{360} = 288\pi \quad ; \quad 36 \cdot \frac{x}{10} = 288 \quad \therefore \ x = 80$$

#8 Find the area of the shaded part in each figure :

(1)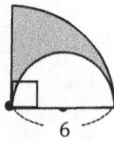

(2)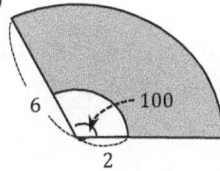

(3)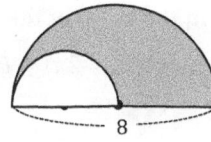

∵ (1) $\frac{1}{4} \cdot \pi \cdot 6^2 - \frac{1}{2} \cdot \pi \cdot 3^2 = 9\pi - \frac{9}{2}\pi = 4\frac{1}{2}\pi$

(2) $\pi \cdot 6^2 \cdot \frac{100}{360} - \pi \cdot 2^2 \cdot \frac{100}{360} = 10\pi - \frac{10}{9}\pi = \frac{80}{9}\pi = 8\frac{8}{9}\pi$

(3) $\frac{1}{2} \cdot \pi \cdot 4^2 - \frac{1}{2} \cdot \pi \cdot 2^2 = 8\pi - 2\pi = 6\pi$

(4)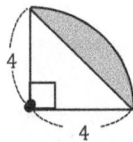

(5)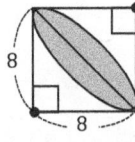

(6)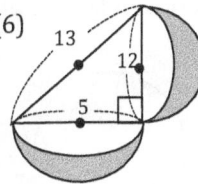

(4) $\frac{1}{4} \cdot \pi \cdot 4^2 - \frac{1}{2} \cdot 4 \cdot 4 = 4\pi - 8$

(5) $2\left(\frac{1}{4} \cdot \pi \cdot 8^2 - \frac{1}{2} \cdot 8 \cdot 8 \right) = 2(16\pi - 32) = 32\pi - 64$

(6) $\frac{1}{2} \cdot \pi \cdot 6^2 + \frac{1}{2} \cdot \pi \cdot \left(\frac{5}{2}\right)^2 + \frac{1}{2} \cdot 5 \cdot 12 - \frac{1}{2} \cdot \pi \cdot \left(\frac{13}{2}\right)^2$

$= 18\pi + \frac{25}{8}\pi + 30 - \frac{169}{8}\pi = \frac{144+25-169}{8}\pi + 30 = 30$

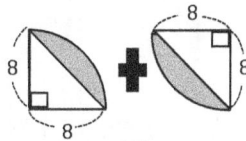

(5)

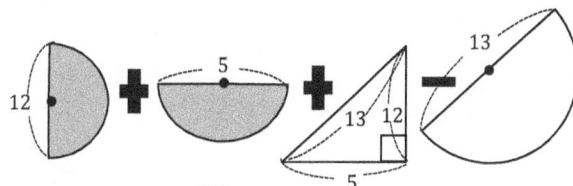

(6)

#9 For two circles with radii r_1, r_2, circumferences C_1, C_2 and areas A_1, A_2, respectively,

the ratio of r_1 to r_2 is $2 : 3$. Answer the following.

(1) Find the ratio of the circumference of the circles.

$\because$ Since $C = 2\pi r$, $C_1 : C_2 = 2\pi r_1 : 2\pi r_2 = r_1 : r_2 = 2 : 3$

(2) Find the ratio of the areas of the circles.

$\because$ Since $A = \pi r^2$, $A_1 : A_2 = \pi r_1^2 : \pi r_2^2 = r_1^2 : r_2^2 = 2^2 : 3^2 = 4 : 9$

#10 $\overset{\frown}{AB} : \overset{\frown}{BC} : \overset{\frown}{AC} = 2 : 3 : 4$

Find the measures of $\angle AOB$ $\angle BOC$, and $\angle AOC$.

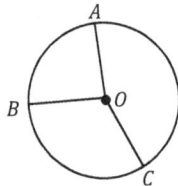

$\because$ $\overset{\frown}{AB} : \overset{\frown}{BC} : \overset{\frown}{AC} = 2 : 3 : 4$ $\Rightarrow$ $m\overset{\frown}{AB} : m\overset{\frown}{BC} : m\overset{\frown}{AC} = 2 : 3 : 4$

$\therefore$ $m\angle AOB : m\angle BOC : m\angle AOC = 2 : 3 : 4$

Therefore, $m\angle AOB = 360 \cdot \dfrac{2}{9} = 80$,

$m\angle BOC = 360 \cdot \dfrac{3}{9} = 120$, and $m\angle AOC = 360 \cdot \dfrac{4}{9} = 160$

#11 Find the ratio of $\overset{\frown}{OB} : \overset{\frown}{AB}$.

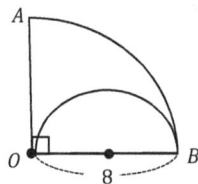

$\because$ $m\overset{\frown}{OB} = \dfrac{1}{2} \cdot 2\pi \cdot 4 = 4\pi$ and $m\overset{\frown}{AB} = \dfrac{1}{4} \cdot 2\pi \cdot 8 = 4\pi$

$\therefore$ $\overset{\frown}{OB} : \overset{\frown}{AB} = 1 : 1$

Solutions for Chapter 7

#1 Find the volume for the following :

(1)

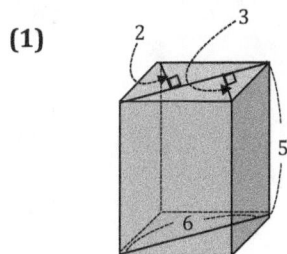

(2)

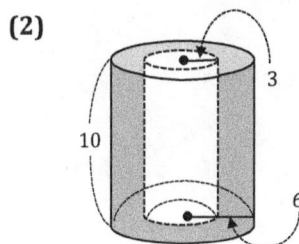

(3) The surface area of the regular prism is 294 in.^{2}

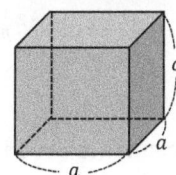

(1) $V = \left(\frac{1}{2} \cdot 6 \cdot 2\right) 5 + \left(\frac{1}{2} \cdot 6 \cdot 3\right) 5 = 30 + 45 = 75$ cubic units

(2) $V = \pi r_1^2 h - \pi r_2^2 h = \pi \cdot 6^2 \cdot 10 - \pi \cdot 3^2 \cdot 10 = 10\pi(36 - 9) = 270\pi$ cubic units

(3) Since $S = 6a^2 = 294$, $a^2 = 49$; $a = 7$ $\quad \therefore V = (7 \cdot 7) \cdot 7 = 343$ in.^{3}

(4)

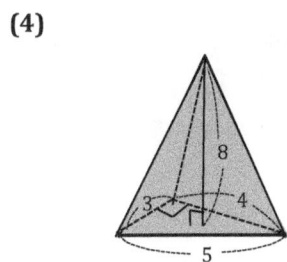

(5)

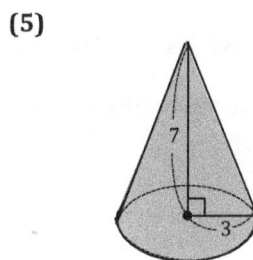

(6)

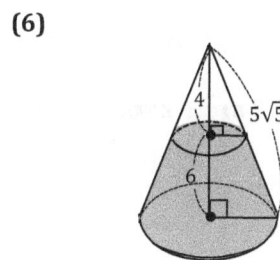

(4) $V = \frac{1}{3}\left(\frac{1}{2} \cdot 3 \cdot 4\right) 8 = \frac{1}{3} \cdot 48 = 16$ cubic units

(5) $V = \frac{1}{3}(\pi r^2 h) = \frac{1}{3}(\pi \cdot 9 \cdot 7) = 21\pi$ cubic units

(6) $V = \frac{1}{3}(\pi r_1^2 h) - \frac{1}{3}(\pi r_2^2 h) = \frac{1}{3}(\pi \cdot 5^2 \cdot 10) - \frac{1}{3}(\pi \cdot 2^2 \cdot 4)$

$\quad = \frac{250}{3}\pi - \frac{16}{3}\pi = \frac{234}{3}\pi = 78\pi$ cubic units

(7)
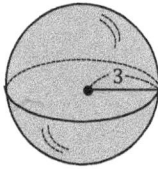

(8)
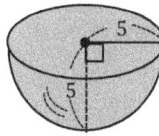

(7) $V = \frac{4}{3}\pi r^3 = \frac{4}{3}\pi \cdot 3^3 = 36\pi$ cubic units

(8) $V = (\frac{4}{3}\pi r^3)\frac{1}{2} = \frac{2}{3}\pi \cdot 5^3 = \frac{250}{3}\pi$ cubic units

#2 Find the surface area for the following :

(1)
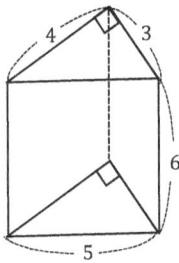

(2)
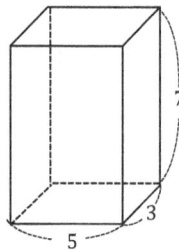

(3)
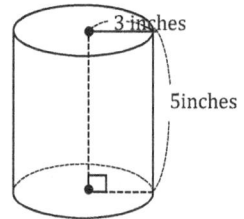

(1) $S = 2 \cdot \left(\frac{1}{2} \cdot 3 \cdot 4\right) + (3 + 4 + 5) \cdot 6 = 12 + 72 = 84$ square units

(2) $S = 2 \cdot (5 \cdot 3) + 2 \cdot (3 \cdot 7) + 2 \cdot (5 \cdot 7) = 30 + 42 + 70 = 142$ square units

(3) $S = 2\pi r^2 + 2\pi rh = 2\pi \cdot 9 + 2\pi \cdot 3 \cdot 5 = 18\pi + 30\pi = 48\pi$ in.^{2}

(4)
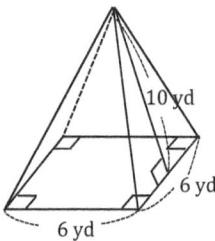

(5)
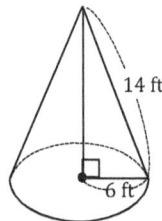

(6)
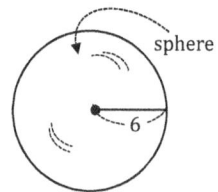

(4) $S = 6 \cdot 6 + 4\left(\frac{1}{2} \cdot 6 \cdot 10\right) = 36 + 120 = 156$ yd^2

(5) $S = \pi r^2 + \frac{1}{2} \cdot 2\pi r \cdot 14 = \pi \cdot 6^2 + \frac{1}{2} \cdot 12\pi \cdot 14 = 36\pi + 84\pi = 120\pi$ ft^2

(6) $S = 4\pi r^2 = 4\pi \cdot 36 = 144\pi$ square units

#3 Find the ratio for the following :

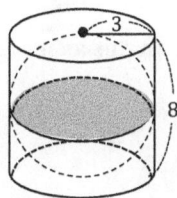

(1) Volume of the cylinder : Volume of the sphere

($\because$ Volume of the cylinder is $\pi r^2 h = \pi \cdot 9 \cdot 8 = 72\pi$

Volume of the sphere is $\frac{4}{3}\pi r^3 = \frac{4}{3}\pi \cdot 3^3 = 36\pi$

$\therefore$ Volume of the cylinder : Volume of the sphere $= 72\pi \ : \ 36\pi = 2 \ : \ 1$)

(2) Surface area of the cylinder : Surface area of the sphere

($\because$ Surface area of the cylinder is $2\pi r^2 + 2\pi r h = 2\pi \cdot 9 + 2\pi \cdot 3 \cdot 8 = 18\pi + 48\pi = 66\pi$

Surface area of the sphere is $4\pi r^2 = 4\pi \cdot 9 = 36\pi$

$\therefore$ Surface area of the cylinder : Surface area of the sphere $= 66\pi : 36\pi = 11 : 6$)

(3) Surface area of the sphere : Area of the great circle of the sphere

($\because$ Surface area of the sphere : Area of the great circle of the sphere $= 4\pi r^2 : \pi r^2 = 4 \ : \ 1$)

Solutions for Chapter 8

#1 Given two similar polygons with lengths of sides or measures of angles as marked, find the specifies lengths/angles :

(1) $\triangle ABC \sim \triangle DEF$ Find a and b.

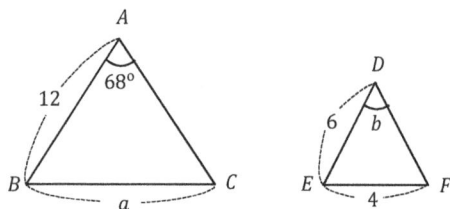

$(\because a = 8 , \ b = 68^\circ)$

(2) $\triangle ABC \sim \triangle DEF$ Find $m \angle D$ and $m \angle F$.

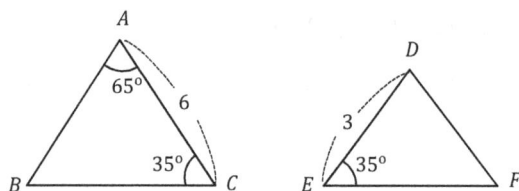

$(\because$ Since $m \angle A \cong m \angle D$, $m \angle D = 65^\circ$

Since $m \angle B = 180^\circ - (65^\circ + 35^\circ) = 80^\circ$ and $m \angle B \cong m \angle F$, $m \angle F = 80^\circ)$

(3) $\square ABCD \sim \square EFGH$ Find AB and $m \angle A$.

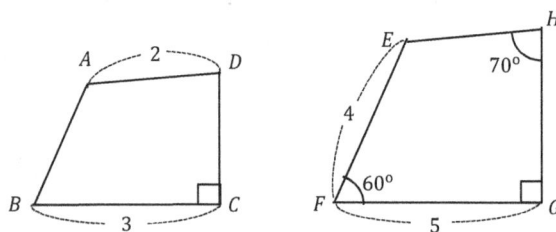

$(\because$ Since $\dfrac{AB}{BC} = \dfrac{EF}{FG}$, $\dfrac{AB}{3} = \dfrac{4}{5}$ $\therefore AB = \dfrac{12}{5}$

Since $\angle A \cong \angle E$ and $m \angle E = 360^\circ - (60^\circ + 90^\circ + 70^\circ) = 140^\circ$, $m \angle A = 140^\circ)$

(4) $\triangle ABC \sim \triangle DEF$ Find a and b.

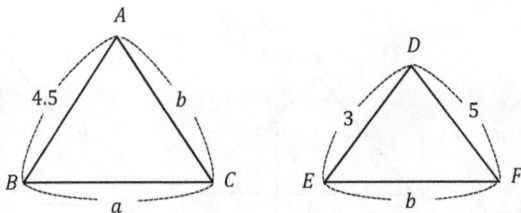

($\because$ Since $\frac{4.5}{3} = \frac{b}{5}$, $3b = 5 \times 4.5$; $b = 7.5$

Since $\frac{4.5}{3} = \frac{a}{b} = \frac{a}{7.5}$, $3a = 4.5 \times 7.5$; $a = \frac{33.75}{3} = 11.25$)

(5) $\square\, ABCD \sim \square\, EFGH$ with $AB : EF = 2 : 3$ Find the perimeter of $\square\, EFGH$.

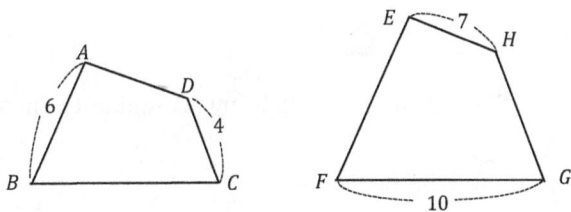

($\because$ Since $AB : EF = 6 : EF = 2 : 3$, $2EF = 18$ $\therefore EF = 9$

Since $DC : HG = 4 : HG = 2 : 3$, $2HG = 12$ $\therefore HG = 6$

Therefore, the perimeter of $\square\, EFGH$ is $9 + 10 + 6 + 7 = 32$.)

(6) $\triangle ABC \sim \triangle EFD$ Find the lengths of $\overline{AB}$ and $\overline{DF}$ and the measure of $\angle C$.

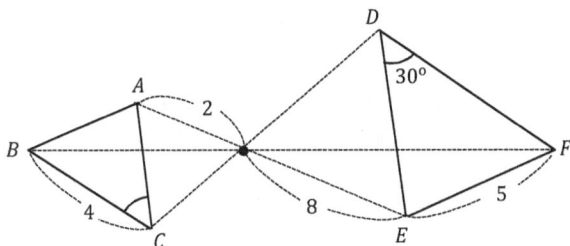

($\because$ $AB : EF = AB : 5 = 1 : 4$ $\therefore AB = \frac{5}{4}$

$BC : DF = 4 : DF = 1 : 4$ $\therefore DF = 16$

Since $\angle C \cong \angle D$, $m \angle C = 30°$)

#2 Find the value of a for the following triangles :

E

(1)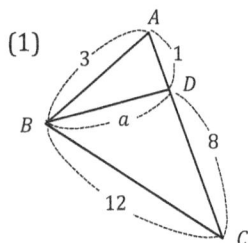

(2)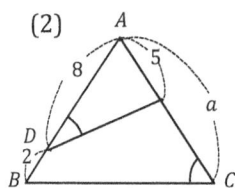

(3)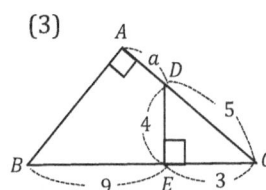

(4)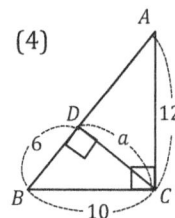

∵

(1)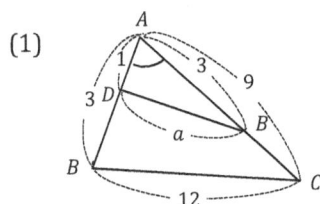

Since $\triangle ADB \sim \triangle ABC$ (by SAS similarity Theorem),

$a : 12 = 1 : 3 \quad \therefore a = 4$

(2)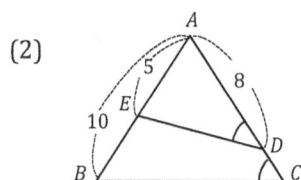

Since $\triangle ABC \sim \triangle AED$ (by AA similarity Theorem),

$\frac{AE}{AB} = \frac{AD}{AC} \ ; \ \frac{5}{10} = \frac{8}{AC}$

$\therefore AC = 16$

(3)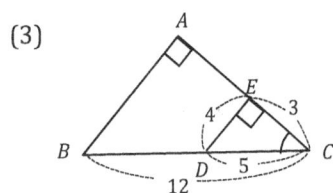

Since $\triangle ABC \sim \triangle EDC$ (by AA similarity Theorem),

$\frac{BC}{DC} = \frac{AC}{EC} \ ; \ \frac{12}{5} = \frac{AC}{3} \quad \therefore AC = \frac{36}{5}$

Since $AC = AD + DC$, $\frac{36}{5} = a + 5$

Therefore, $a = \frac{36}{5} - 5 = \frac{11}{5}$

(4)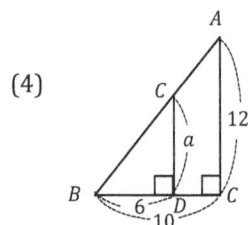

$\frac{6}{a} = \frac{10}{12} \quad \therefore 10a = 6 \cdot 12$

Therefore, $a = \frac{36}{5}$

(5)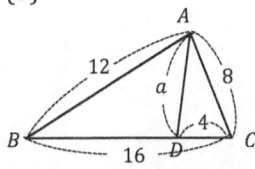

(6)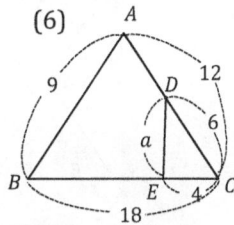

(7)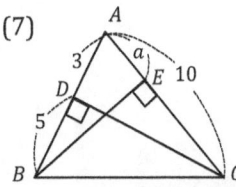

(8)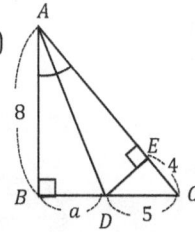

∵

(5)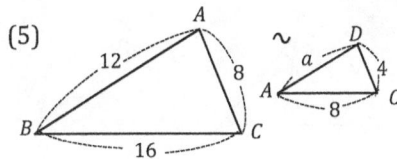

Since $AC : BC = 1 : 2$, $DC : AC = 1 : 2$, and $\angle C$ is common,

$\triangle ABC \sim \triangle DAC$

∴ $\dfrac{12}{a} = \dfrac{8}{4}$, $\ a = \dfrac{12 \cdot 4}{8} = 6$

(6)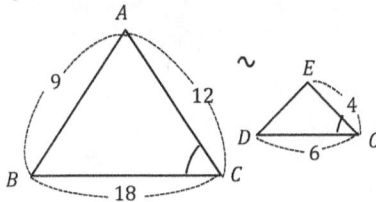

Since $AC : EC = 3 : 1$, $BC : DC = 3 : 1$, and $\angle C$ is common,

$\triangle ABC \sim \triangle EDC$

∴ $AB : ED = 9 : ED = 3 : 1$

Therefore, $a = ED = 3$

(7)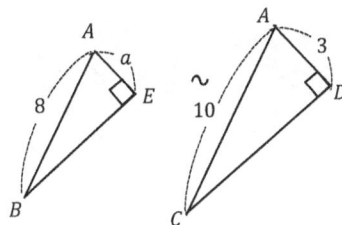

$\triangle ABE \sim \triangle ACD$

∴ $8 : 10 = a : 3$

Therefore, $a = \dfrac{3 \cdot 8}{10} = \dfrac{12}{5}$

(8)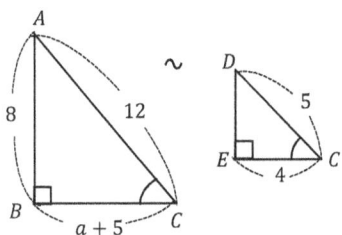

Since $\triangle ABD \cong \triangle AED$, $AE = 8$

∴ $AC = 8 + 4 = 12$

Since $\triangle ABC \sim \triangle DEC$, $12 : 5 = a + 5 : 4$

∴ $5(a + 5) = 12 \cdot 4$; $a + 5 = \dfrac{48}{5}$

Therefore, $a = \dfrac{48}{5} - 5 = \dfrac{23}{5}$

#3 $\overline{DE} /\!/ \overline{BC}$. **Find the value of** a **and** b **for the following triangles :**

(1)

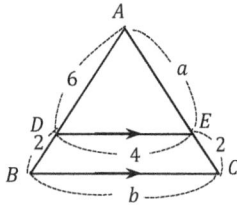

(2)

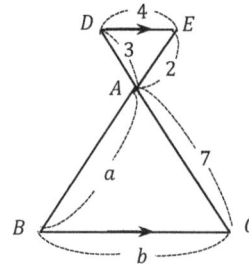

(3)

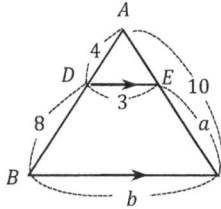

(4)

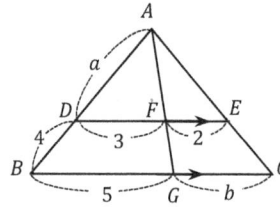

$\because$ (1) $6 : a = 8 : a + 2$; $8a = 6(a+2) = 6a + 12$ $\therefore 2a = 12$; $a = 6$

$6 : 4 = 8 : b$; $6b = 4 \cdot 8$ $\therefore b = \dfrac{4 \cdot 8}{6} = \dfrac{16}{3}$

(2) $a : 2 = 7 : 3$; $3a = 14$ $\therefore a = \dfrac{14}{3}$

$4 : b = 3 : 7$; $3b = 4 \cdot 7$ $\therefore b = \dfrac{28}{3}$

(3) $4 : 3 = 12 : b$; $4b = 3 \cdot 12$ $\therefore b = 9$

$12 : 8 = 10 : a$; $12a = 8 \cdot 10$ $\therefore a = \dfrac{20}{3}$

(4) $a : a + 4 = 3 : 5$; $5a = 3(a+4)$; $2a = 12$ $\therefore a = 6$

$AF : AG = DF : BG = FE : GC$ $\therefore$ $3 : 5 = 2 : b$; $3b = 10$ $\therefore b = \dfrac{10}{3}$

#4 For each triangle ΔABC, $\overline{AD}$ **is the bisector of** $\angle A$. **Find the value of** a.

(1)

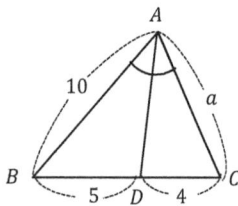

(2)

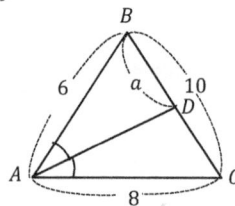

(3)

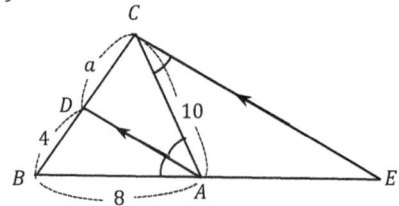

∵ (1) $10 : a = 5 : 4$ (by the angle bisector theorem) ∴ $5a = 10 \cdot 4$; $a = 8$

(2) $8 : 6 = (10 - a) : a$ (by the angle bisector theorem)

$$8a = 6(10 - a) = 60 - 6a \; ; \; 14a = 60 \quad \therefore \quad a = \frac{60}{14} = \frac{30}{7}$$

(3) $\angle DAC \cong \angle ACE$ by alternate interior angles.

$\angle BAD \cong \angle AEC$ by corresponding angles. ∴ $\angle ACE \cong \angle AEC$

∴ ΔAEC is isosceles. ; $AC = AE = 10$

∴ $8 : 4 = 10 : a$; $8a = 40$; $a = 5$

#5 For each triangle ΔABC, $\overline{AD}$ is the bisector of an exterior angle at A. Find the value of a.

(1)

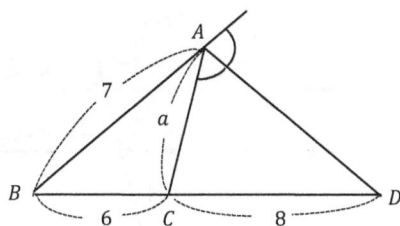

(2)

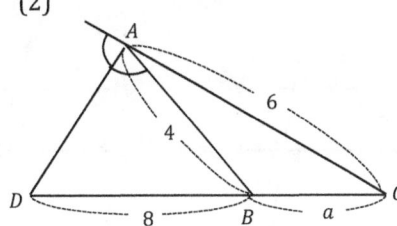

∵ (1) $7 : a = 14 : 8$ (by the angle bisector theorem). So, $14a = 7 \cdot 8$ ∴ $a = 4$

(2) $6 : 4 = 8 + a : 8$ (by the angle bisector theorem). So, $4(8 + a) = 8 \cdot 6$ ∴ $a = 4$

#6 Answer the following :

(1)

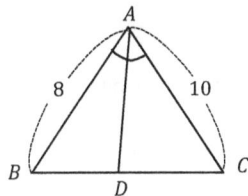

In triangle ΔABC, $\overline{AD}$ is the bisector of $\angle A$.

The area of ΔABD is 12 square units.

Find the area of ΔADC.

∵ Since $AB : AC = BD : DC$ (by the angle bisector theorem), $8 : 10 = 4 : 5 = BD : DC$

(The area of ΔABD : Tha area of ΔADC) = (12 : Tha area of ΔADC) = $4 : 5$

Therefore, Tha area of ΔADC is $\frac{5 \cdot 12}{4} = 15$ square units.

(2)

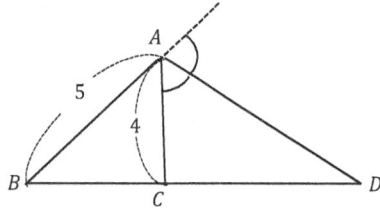

In triangle $\triangle ABC$,

$\overline{AD}$ is the bisector of an exterior angle at A.

Find the ratio of the areas of $\triangle ABD$ and $\triangle ACD$.

∵ Since $AB : AC = BD : CD$ (by the angle bisector theorem), $5 : 4 = BD : CD$

∴ The area of $\triangle ABD$: Tha area of $\triangle ACD$

$= \frac{1}{2} \cdot BD \cdot \text{height} : \frac{1}{2} \cdot CD \cdot \text{height}$

$= BD : CD$

$= 5 : 4$

#7 $l \mathbin{/\!/} m \mathbin{/\!/} n$ Find the value of a.

(1)

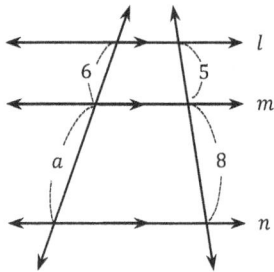

(2)

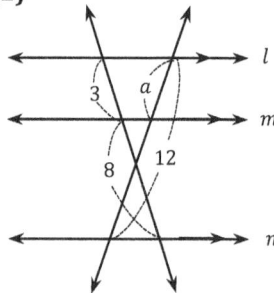

(3)

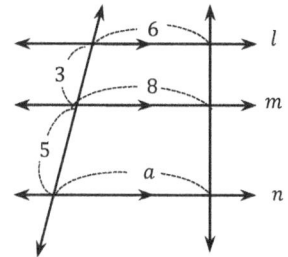

∵ (1) $6 : a = 5 : 8$ ∴ $5a = 48$; $a = \dfrac{48}{5}$

(2) $3 : 8 = a : 12 - a$; $8a = 3(12 - a)$; $11a = 36$ ∴ $a = \dfrac{36}{11}$

(3)

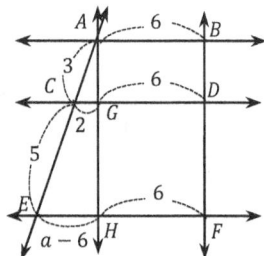

Consider $\overline{AH} \mathbin{/\!/} \overline{BF}$

Then, $3 : 2 = 8 : a - 6$; $3(a - 6) = 16$

Therefore, $3a = 34$ ∴ $a = \dfrac{34}{3}$

(4)

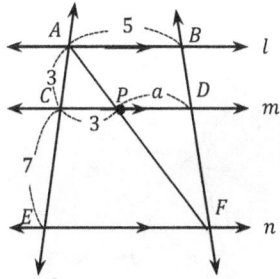

(5)

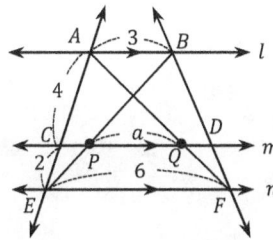

(6)

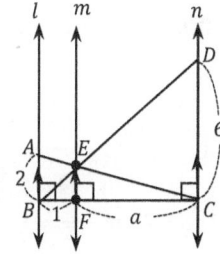

∵ (4) Since $CP : EF = 3 : 3 + 7$, $PD : 5 = 7 : 3 + 7$, and $CD = CP + PD$, $CD = \frac{3 \cdot EF + 7 \cdot 5}{3 + 7}$

Since $AC : CP = AE : EF$, $3 : 3 = 10 : EF$ ∴ $EF = 10$

Therefore, $CD = \frac{3 \cdot 10 + 7 \cdot 5}{3 + 7} = \frac{65}{10} = 6\frac{1}{2}$

∴ $a = 6\frac{1}{2} - 3 = 3\frac{1}{2}$

(5) $4 : 6 = CQ : 6$ ∴ $CQ = 4$

$CP : 3 = 2 : 6 = 1 : 3$ ∴ $CP = 1$

Therefore, $a = PQ = CQ - CP = 4 - 1 = 3$

(6) $EF : 2 = a : a + 1$; $2a = (a + 1)EF$

$EF : 6 = 1 : a + 1$; $(a + 1)EF = 6$

Therefore, $2a = 6$; $a = 3$

#8 For triangles $\triangle ABC$, answer the questions :

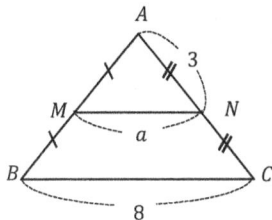

(1) M and N are midpoints of $\overline{AB}$ and $\overline{AC}$, respectively. Find the value of a and the length of $\overline{AC}$.

(∵ $a = \frac{1}{2} \cdot 8 = 4$ and

$AC = AN + NC = 3 + 3 = 6$)

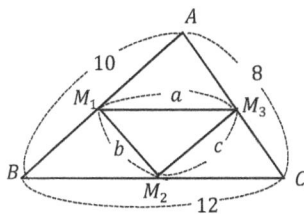

(2) M_1, M_2 and M_3 are midpoints of $\overline{AB}$, $\overline{BC}$, and $\overline{AC}$, respectively. Find the value of $a + b + c$.

(∵ $a = \frac{12}{2} = 6$, $b = \frac{8}{2} = 4$, and $c = \frac{10}{2} = 5$

∴ $a + b + c = 6 + 4 + 5 = 15$)

#9 For parallelograms $\square\, ABCD$ with $\overline{AD}\,/\!/\,\overline{BC}$, M and N are midpoints of $\overline{AB}$ and $\overline{DC}$. Find the value of a.

(1) 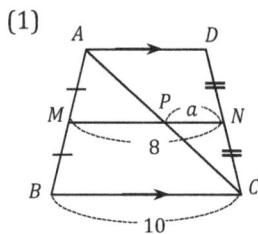 (2) 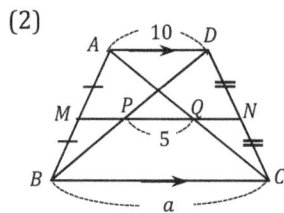 (3)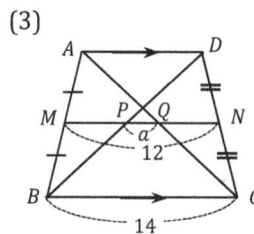

$\because$ (1) $MP = \frac{1}{2} \cdot BC = \frac{1}{2} \cdot 10 = 5$ $\therefore a = 8 - 5 = 3$

(2) $MP = \frac{1}{2} \cdot AD = \frac{1}{2} \cdot 10 = 5$ $\therefore a = BC = 2 \cdot MQ = 2(5 + 5) = 20$

(3) $MQ = \frac{BC}{2} = \frac{14}{2} = 7$

Since $MN = 12$, $QN = 5$

Since $PN = \frac{BC}{2} = 7$, $a = PQ = 2$

#10 The point P is the centroid of triangles $\triangle\, ABC$. Find the values of a and b.

(1) 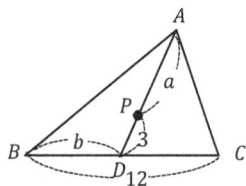 (2)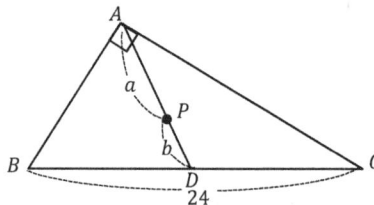

$\because$ (1) Since $BD = DC$, $b = 6$

Since $a : 3 = 2 : 1$, $a = 6$

(2) Since $AD = 12$ and $a : b = 2 : 1$, $a = 12 \cdot \frac{2}{3} = 8$ and $b = 12 \cdot \frac{1}{3} = 4$

#11 Points P and Q are the centroids of the triangles $\triangle ABC$ and $\triangle PBC$. Answer the questions.

(1)

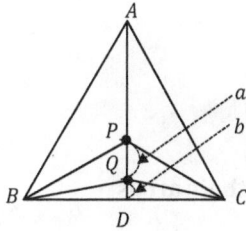

$AD = 24$.

Find the lengths of a and b, where $a = PQ, b = QD$.

($\because$ Since $PD = \frac{1}{3} \cdot 24 = 8$,

$a = 8 \cdot \frac{2}{3} = \frac{16}{3}$ and $b = QD = 8 \cdot \frac{1}{3} = \frac{8}{3}$)

(2)

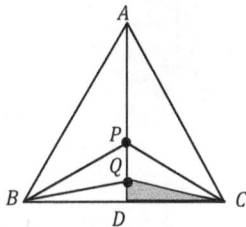

The area of $\triangle QDC$ is 5.

Find the area of $\triangle ABC$.

($\because$ Since $PQ : QD = 2 : 1$, the area of $\triangle PQC$ is 10.

$\therefore$ The area of $\triangle PDC$ is 15.

$\therefore$ The area of $\triangle ABC = 6 \cdot 15 = 90$ square units.)

#12 For parallelograms $\square ABCD$ with areas of 72 in.^{2}, answer the following :

(1)

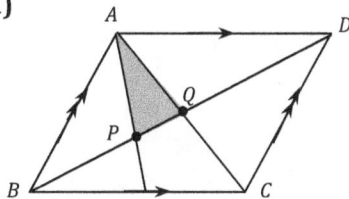

P is the centroid of $\triangle ABC$ and

Q is the intersection point of two diagonals of $\square ABCD$.

Find the area of $\triangle APQ$.

($\because$ Since P is the centroid of $\triangle ABC$, the area of $\triangle APQ$ is $\frac{1}{6} \cdot$ (the area of $\triangle ABC$)

Since the area of $\triangle ABC$ is $\frac{1}{2} \cdot$ (the area of $\square ABCD$),

the area of $\triangle APQ = \frac{1}{6} \cdot \left(\frac{1}{2} \cdot 72 \right) = \frac{1}{12} \cdot 72 = 6$ in.^{2})

(2)

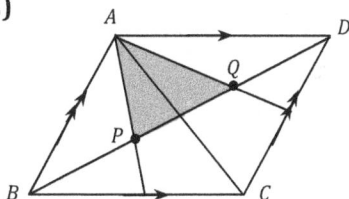

P is the centroid of $\triangle ABC$ and

Q is the centroid of $\triangle ACD$.

Find the area of $\triangle APQ$.

($\because$ Since $BP = PQ = QD$, the area of $\triangle APQ$ is $\frac{1}{3} \cdot$ (the area of $\triangle ABD$).

Since the area of $\triangle ABD$ is $\frac{1}{2} \cdot$ (the area of $\square ABCD$),

the area of $\triangle APQ = \frac{1}{6} \cdot 72 = 12$ in.^{2})

#13 For two similar polygons in each question, find the ratio of their areas.

(1)
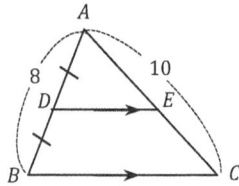

(2)
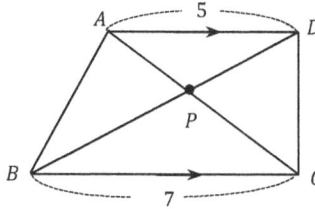

($\because$ (1)   Since $\triangle ADE \sim \triangle ABC$ and $AD : AB = 1 : 2$,

the area of $\triangle ADE$: the area of $\triangle ABC = 1^2 : 2^2 = 1 : 4$)

(2)   Since $\triangle ADP \sim \triangle CBP$ and $AD : BC = 5 : 7$,

the area of $\triangle ADP$: the area of $\triangle CBP = 5^2 : 7^2 = 25 : 49$)

#14 Answer the following :

(1)
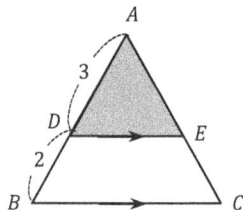

For triangle $\triangle ABC$,

$\overline{DE} \ // \ \overline{BC}$ and the area of $\triangle ADE$ is 30 in.^{2}

Find the area of $\square DBCE$.

($\because$ Since $\triangle ADE \sim \triangle ABC$ and $AD : AB = 3 : 5$,

the area of $\triangle ADE$: the area of $\triangle ABC = 3^2 : 5^2 = 9 : 25$

So,   30 : the area of $\triangle ABC = 9 : 25$

$\therefore$ The area of $\triangle ABC$ is $\frac{30 \cdot 25}{9} = \frac{250}{3}$ and

The area of $\square DBCE$ is $\frac{250}{3} - 30 = \frac{160}{3}$ square units.)

(2)

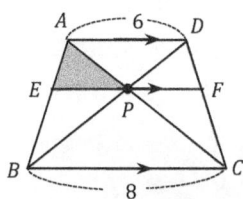

For trapezoid ▱ $ABCD$,
$\overline{AD} \parallel \overline{EF} \parallel \overline{BC}$.

The area of $\triangle AEP$ is 10 in.^{2}

Find the area of $\triangle ABC$.

(∵ Since $\triangle ADP \sim \triangle CBP$ and $AD : BC = 6 : 8 = 3 : 4$, $AP : PC = 3 : 4$

∴ $AP : AC = 3 : 7$

∴ the area of $\triangle AEP$: the area of $\triangle ABC = 3^2 : 7^2$

∴ 10 : the area of $\triangle ABC = 9 : 49$

∴ The area of $\triangle ABC$ is $\dfrac{10 \cdot 49}{9} = \dfrac{490}{9}$ in.^{2})

#15 **For two similar cones A and B, $h_1 : h_2 = 3 : 5$. Answer the following :**

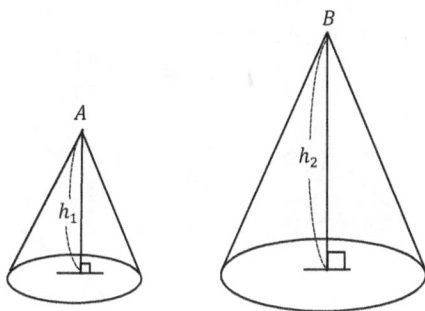

(1) The area S_1 of the lateral face of A is 144 in.^{2} Find the area S_2 of the lateral face of B.

(∵ Since $S_1 : S_2 = 3^2 : 5^2$, $144 : S_2 = 3^2 : 5^2$

∴ $12^2 : S_2 = 3^2 : 5^2$

∴ $S_2 = 20^2 = 400$ in.^{2})

(2) The volume V_1 of A is 216 in.^{3} Find the volume V_2 of B.

(∵ Since $V_1 : V_2 = 3^3 : 5^3$, $216 : V_2 = 3^3 : 5^3$

∴ $6^3 : V_2 = 3^3 : 5^3$

∴ $V_2 = 10^3 = 1000$ in.^{3})

Solutions for Chapter 9

#1 Find the value of x.

(1)

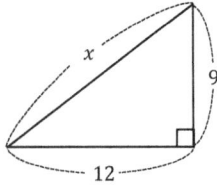

(2)

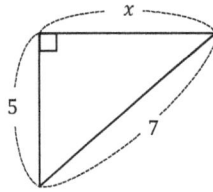

(3)

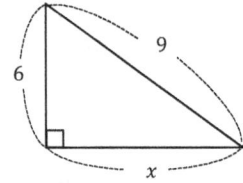

$$\therefore (1)\ x^2 = 12^2 + 9^2 = 144 + 81 = 225 = 15^2 \ \therefore\ x = 15$$

$$(2)\ 7^2 = 5^2 + x^2\ ;\ \ x^2 = 49 - 25 = 24 = 3 \cdot 8 = 3 \cdot 2^3 \ \therefore\ x = 2\sqrt{6}$$

$$(3)\ 9^2 = 6^2 + x^2\ ;\ \ x^2 = 81 - 36 = 45 = 5 \cdot 9 = 5 \cdot 3^2 \ \therefore\ x = 3\sqrt{5}$$

(4)

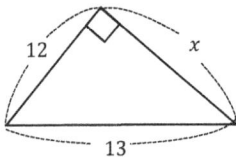

(5)

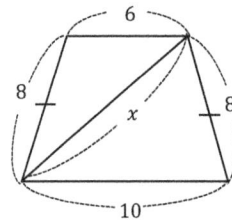

(6)

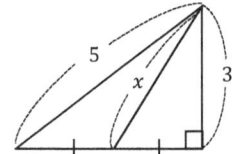

$$\therefore (4)\ 13^2 = 12^2 + x^2\ ;\ x^2 = 169 - 144 = 25 = 5^2 \ \therefore\ x = 5$$

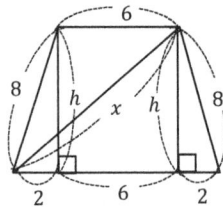

$$(5)\ 8^2 = h^2 + 2^2\ ;\ \ h^2 = 64 - 4 = 60$$

$$x^2 = h^2 + (2 + 6)^2 = 60 + 64 = 124 \ \ \ \therefore\ x = 2\sqrt{31}$$

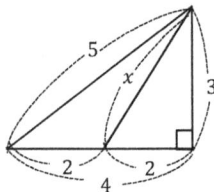

$$(6)\ x^2 = 2^2 + 3^2 = 4 + 9 = 13 \ \ \ \therefore\ x = \sqrt{13}$$

#2 Find the values of x and y.

(1)

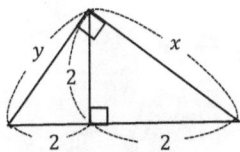

(2)

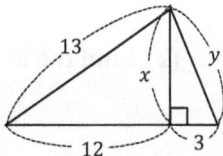

(3)

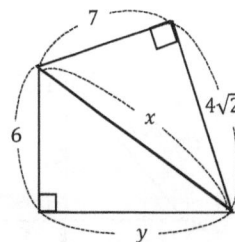

$\therefore$ (1) $x^2 = 2^2 + 2^2 = 4 + 4 = 8 = 2^3$ $\therefore$ $x = 2\sqrt{2}$

$(2+2)^2 = y^2 + x^2 = y^2 + 8$; $y^2 = 16 - 8 = 8$ $\therefore$ $y = 2\sqrt{2}$

(2) $13^2 = 12^2 + x^2$; $x^2 = 169 - 144 = 25 = 5^2$ $\therefore$ $x = 5$

$y^2 = x^2 + 3^2 = 25 + 9 = 34$ $\therefore$ $y = \sqrt{34}$

(3) $x^2 = 7^2 + (4\sqrt{2})^2 = 49 + 16 \cdot 2 = 81 = 9^2$ $\therefore$ $x = 9$

$y^2 = x^2 - 6^2 = 81 - 36 = 45$ $\therefore$ $y = 3\sqrt{5}$

(4)

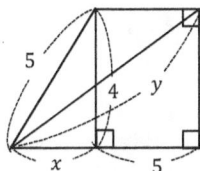

(4) $5^2 = x^2 + 4^2$; $x^2 = 25 - 16 = 9 = 3^2$ $\therefore$ $x = 3$

$y^2 = (3+5)^2 + 4^2 = 64 + 16 = 80 = 16 \cdot 5$ $\therefore$ $y = 4\sqrt{5}$

#3 Find the areas for the following :

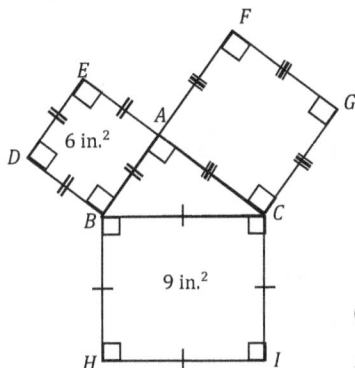

(1) The area of $\square BDEA$ is 6 in.^{2} and the area of $\square BCIH$ is 9 in.^{2} Find the area of $\square ACGF$.

($\because$ Since the area of $\square BCIH$ is the sum of the areas of $\square BDEA$ and $\square ACGF$, the area of $\square ACGF$ is $9 - 6 = 3$ in.^{2})

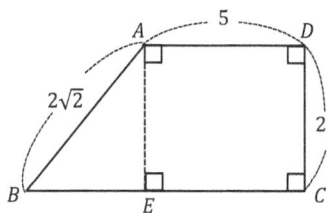

(2) Find the area of $\square ABCD$.

$\because (2\sqrt{2})^2 = AE^2 + BE^2 = 2^2 + BE^2$; $BE^2 = 8 - 4 = 4 = 2^2$ $\therefore BE = 2$

Therefore, the area of $\square ABCD$ is $\frac{1}{2}(5 + 7) \cdot 2 = 12$ square units.

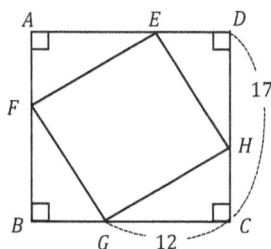

(3) All four right triangles are congruent.
Find the area of $\square EFGH$.

$\because GH^2 = 12^2 + (17 - 12)^2 = 144 + 25 = 169 = 13^2$ $\therefore GH = 13$

Therefore, the area of $\square EFGH$ is $13^2 = 169$ square units.

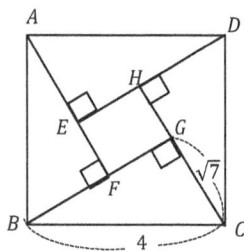

(4) Find the area of $\square EFGH$.

$\because BG^2 = 4^2 - (\sqrt{7})^2 = 16 - 7 = 9 = 3^2$ $\therefore BG = 3$ So, $EF = 3 - \sqrt{7}$

Therefore, the area of $\square EFGH$ is

$EF^2 = \left(3 - \sqrt{7}\right)^2 = 9 - 6\sqrt{7} + 7 = 16 - 6\sqrt{7}$ square units.

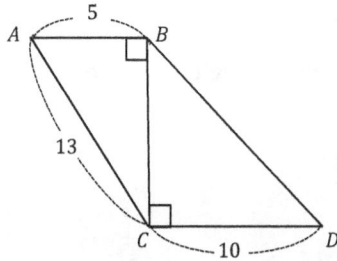

(5) Find the area of △BCD.

$\because BC^2 = 13^2 - 5^2 = 169 - 25 = 144 = 12^2$ $\therefore$ $BC = 12$

Therefore, the area of △BCD is $\frac{1}{2} \cdot 10 \cdot 12 = 60$ square units.

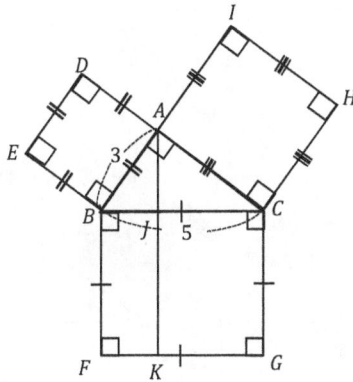

(6) Find the area of □$JKGC$.

$\because$ Since $AC^2 = 5^2 - 3^2 = 25 - 9 = 16 = 4^2$, the area of □$ACHI$ is 16.

Since the area of □$ACHI$ is equal to the area of □$JKGC$,

the area of □$JKGC$ is 16 square units.

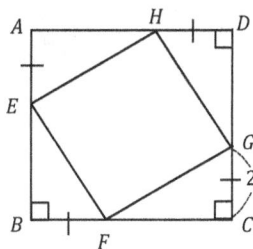

(7) The area of □$EFGH$ is 18 in.²
Find the area of □$ABCD$.

$\because$ Since $FG = \sqrt{18} = 3\sqrt{2}$, $FC^2 = FG^2 - 2^2 = 18 - 4 = 14$

$\therefore$ $FC = \sqrt{14}$; $BC = \sqrt{14} + 2$

Therefore, the area of □$ABCD$ is

$(\sqrt{14} + 2)^2 = 14 + 4\sqrt{14} + 4 = 18 + 4\sqrt{14}$ in.²

#4 Determine whether triangles with the following side lengths are right triangles.

(1) **2, 2, 4** ; $4^2 \neq 2^2 + 2^2$ ∴ not a right triangle.

(2) **3, 4, 5** ; $5^2 = 3^2 + 4^2$ ∴ a right triangle.

(3) **1, 2, $\sqrt{3}$** ; $2^2 = 1^2 + (\sqrt{3})^2$ ∴ a right triangle.

(4) **7, 9, 11** ; $11^2 \neq 7^2 + 9^2$ ∴ not a right triangle.

(5) **$\sqrt{2}$, $\sqrt{3}$, $\sqrt{5}$** ; $(\sqrt{5})^2 = (\sqrt{2})^2 + (\sqrt{3})^2$ ∴ a right triangle.

(6) **5, 12, 13** ; $13^2 = 12^2 + 5^2$ ∴ a right triangle.

#5 The triangles ΔABC are right triangles with $m \angle C = 90°$. Find the value of x.

(1) (2) (3) (4)

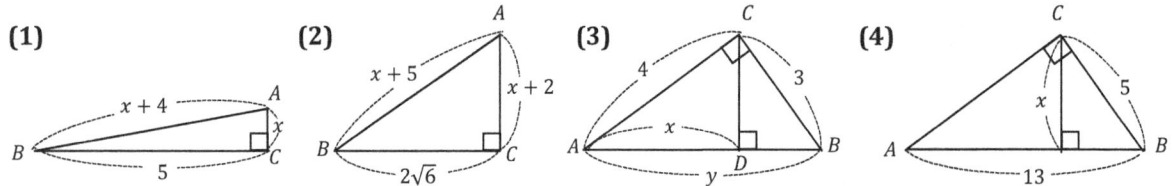

∴ (1) $(x + 4)^2 = 5^2 + x^2$; $x^2 + 8x + 16 = x^2 + 25$; $8x = 9$ ∴ $x = \frac{9}{8}$

(2) $(x + 5)^2 = (2\sqrt{6})^2 + (x + 2)^2$; $x^2 + 10x + 25 = 24 + x^2 + 4x + 4$; $6x = 3$ ∴ $x = \frac{1}{2}$

(3) $y^2 = 3^2 + 4^2 = 9 + 16 = 25 = 5^2$ ∴ $y = 5$

$4^2 = x \cdot y = x \cdot 5$ ∴ $x = \frac{16}{5}$

(4) $AC^2 = 13^2 - 5^2 = 12^2$; $12 \cdot 5 = x \cdot 13$ ∴ $x = \frac{60}{13}$

#6 Answer the questions for each figure.

(1)

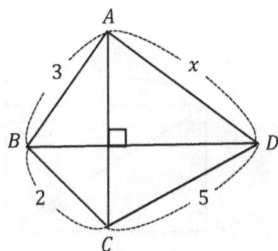

$\overline{AC} \perp \overline{BD}$. Find AD .

(2)

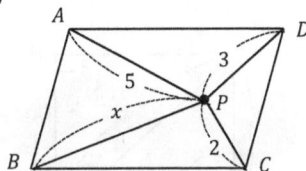

P is an interior point in
$\square\, ABCD$. Find BP .

(3)

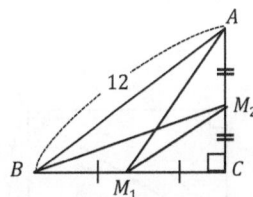

M_1 and M_2 are midpoints
of $\overline{BC}$ and $\overline{AC}$,
respectively.

Find $AM_1{}^2 + BM_2{}^2$.

∵ (1) $3^2 + 5^2 = x^2 + 2^2$; $9 + 25 = x^2 + 4$; $x^2 = 34 - 4 = 30$ ∴ $x = \sqrt{30}$

(2) $5^2 + 2^2 = x^2 + 3^2$; $25 + 4 = x^2 + 9$; $x^2 = 20$ ∴ $x = \sqrt{20} = 2\sqrt{5}$

(3) Since $AB = 12$, $M_1 M_2 = \frac{12}{2} = 6.$ ∴ $AM_1{}^2 + BM_2{}^2 = 6^2 + 12^2 = 36 + 144 = 180$

(4)

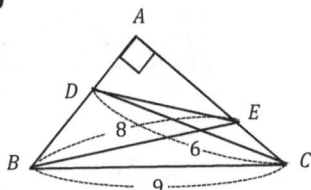

For a right triangle ΔABC
with $m\angle A = 90^\circ$,
$BE = 8, CD = 6$, and $BC = 9$.
Find DE.

(5)

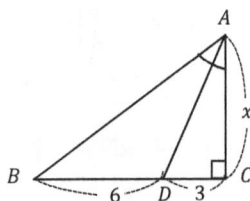

$\angle BAD \cong \angle DAC$
Find the value of x.

(6)

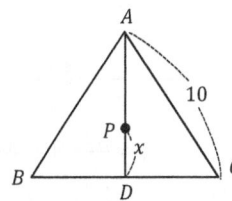

P is the centroid of
a regular triangle ΔABC.
Find the value of x .

∵ (4) $DE^2 + 9^2 = 8^2 + 6^2$ (by the applications of the Pythagorean Theorem)

$DE^2 = 64 + 36 - 81 = 19$ ∴ $DE = \sqrt{19}$

(5) Since $\frac{AB}{x} = \frac{6}{3}$, $AB = 2x$.

$AB^2 = x^2 + (6+3)^2$; $4x^2 = x^2 + 81$; $3x^2 = 81$; $x^2 = 27$ ∴ $x = 3\sqrt{3}$

(6) $AD = \frac{\sqrt{3}}{2} \cdot 10 = 5\sqrt{3}$.

Since $PD = \frac{1}{3} AD$, $x = \frac{1}{3} \cdot 5\sqrt{3} = \frac{5\sqrt{3}}{3}$.

#7 Find the value of x.

(1)

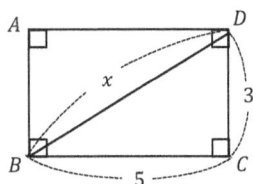

(2)

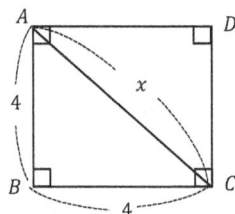

(3)

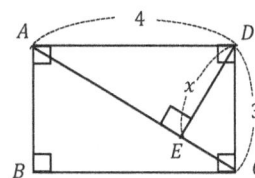

$\because$ (1) $x = \sqrt{5^2 + 3^2} = \sqrt{25 + 9} = \sqrt{34}$

(2) $x = \sqrt{4^2 + 4^2} = \sqrt{32} = 4\sqrt{2}$

(3) $AC^2 = 3^2 + 4^2 = 9 + 16 = 25$ $\therefore AC = 5$

Since $3 \cdot 4 = x \cdot 5, \quad x = \frac{12}{5}$

(4)

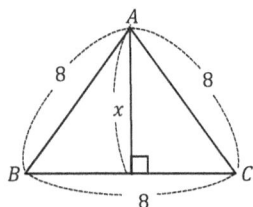

(5)

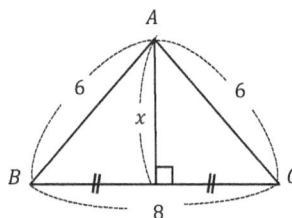

(6)

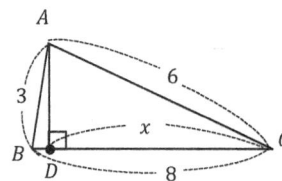

$\because$ (4) $x = \frac{\sqrt{3}}{2} \cdot 8 = 4\sqrt{3}$

(5) $x^2 = 6^2 - 4^2 = 36 - 16 = 20$ $\therefore x = 2\sqrt{5}$

(6) Since $AD^2 = 3^2 - (8 - x)^2$ and $AD^2 = 6^2 - x^2, \quad 9 - (64 - 16x + x^2) = 36 - x^2$

; $16x = 91$ $\therefore x = \frac{91}{16}$

#8 Find the areas of $\triangle ABC$.

(1)

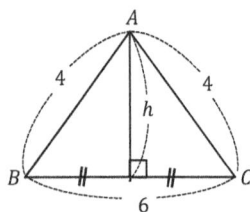

(2)

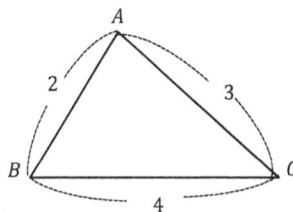

(3)

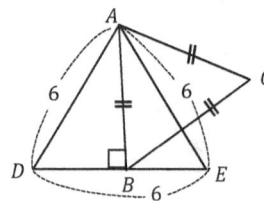

∵ (1) $h^2 = 4^2 - 3^2 = 16 - 9 = 7$ ∴ $h = \sqrt{7}$ Therefore, the area of $\triangle ABC$ is $\frac{1}{2} \cdot 6 \cdot \sqrt{7} = 3\sqrt{7}$

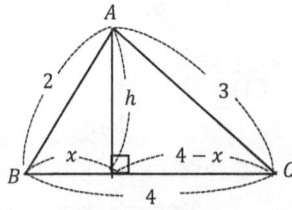

(2) $h^2 = 2^2 - x^2 = 3^2 - (4-x)^2$; $4 - x^2 = 9 - (16 - 8x + x^2)$

$4 = -7 + 8x$; $8x = 11$; $x = \frac{11}{8}$

∴ $h^2 = 4 - \frac{121}{64} = \frac{135}{64}$; $h = \frac{\sqrt{135}}{8} = \frac{3\sqrt{15}}{8}$

Therefore, the area of $\triangle ABC$ is $\frac{1}{2} \cdot 4 \cdot \frac{3\sqrt{15}}{8} = \frac{3\sqrt{15}}{4}$

(3) $AB = \frac{\sqrt{3}}{2} \cdot 6 = 3\sqrt{3}$

Therefore, the area of $\triangle ABC$ is $\frac{\sqrt{3}}{4} \cdot \left(3\sqrt{3}\right)^2 = \frac{\sqrt{3}}{4} \cdot 27 = \frac{27\sqrt{3}}{4}$

#9 Find the values of x and y.

(1)

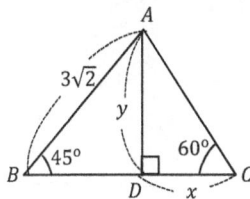

(2)

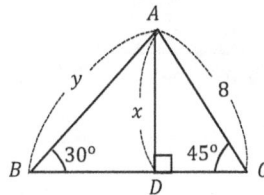

∵ (1)

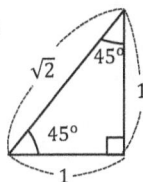

$3\sqrt{2} : y = \sqrt{2} : 1$

∴ $y = 3$

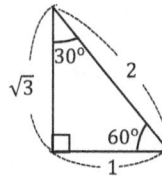

$y : x = \sqrt{3} : 1$

So, $3 : x = \sqrt{3} : 1$

∴ $x = \frac{3}{\sqrt{3}} = \frac{3\sqrt{3}}{3} = \sqrt{3}$

(2)

x value obtained from the right side of this solution.

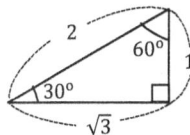

$y : 4\sqrt{2} = 2 : 1$

∴ $y = 8\sqrt{2}$

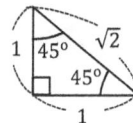

$x : 8 = 1 : \sqrt{2}$

∴ $x = \frac{8}{\sqrt{2}} = \frac{8\sqrt{2}}{2} = 4\sqrt{2}$

#10 Find the value of x.

(1)

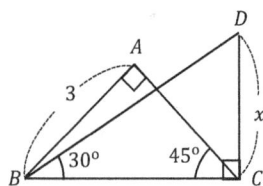

(2)

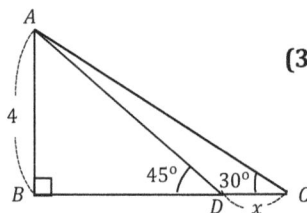

(3)

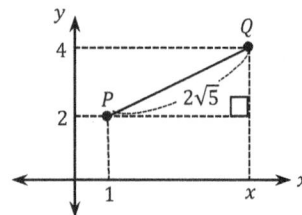

∵ (1) $3 : BC = 1 : \sqrt{2}$ ∴ $BC = 3\sqrt{2}$

So, $x : BC = 1 : \sqrt{3}$ ⇒ $x : 3\sqrt{2} = 1 : \sqrt{3}$ ∴ $x = \frac{3\sqrt{2}}{\sqrt{3}} = \frac{3\sqrt{6}}{3} = \sqrt{6}$

(2) $AB : BD = 1 : 1$ ∴ $BD = 4$

So, $AB : BC = 1 : \sqrt{3}$ ⇒ $4 : 4 + x = 1 : \sqrt{3}$; $4 + x = 4\sqrt{3}$ ∴ $x = 4\sqrt{3} - 4$

(3) $PQ = \sqrt{(x-1)^2 + (4-2)^2} = \sqrt{(x-1)^2 + 4} = 2\sqrt{5}$

∴ $(x-1)^2 + 4 = 20$; $(x-1)^2 = 16 = 4^2$; $x - 1 = 4$ ∴ $x = 5$

(4)

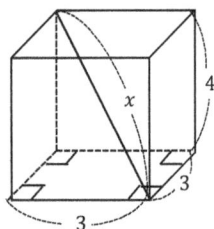

(5)

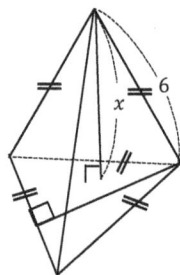

(6)

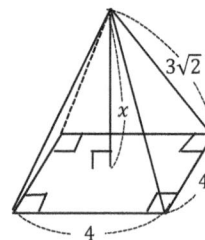

∵ (4) $x = \sqrt{3^2 + 3^2 + 4^2} = \sqrt{9 + 9 + 16} = \sqrt{34}$

(5) $x = \frac{\sqrt{6}}{3} \cdot 6 = 2\sqrt{6}$

(6) $x = \sqrt{(3\sqrt{2})^2 - \frac{4^2}{2}} = \sqrt{18 - \frac{16}{2}} = \sqrt{10}$

See Chapter 9, 9-2, 3. Solids (for more information)!

#11   Answer the questions for each figure.

(1)

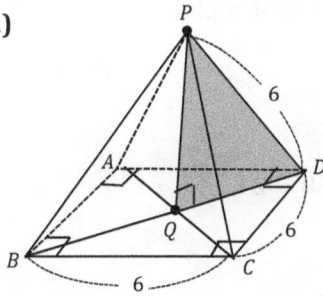

Find the area of Δ *PQD*.

(∵   (1) $BD = 6\sqrt{2}$, $QD = 3\sqrt{2}$

∴ $PQ^2 = 6^2 - \left(3\sqrt{2}\right)^2 = 36 - 18 = 18$ ∴ $PQ = 3\sqrt{2}$

Therefore, the area of Δ *PQD* is $\frac{1}{2} \cdot 3\sqrt{2} \cdot 3\sqrt{2} = 9$)

(2)

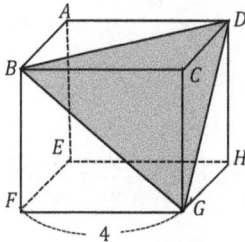

For a cube with sides of length 4, find the area of Δ *BDG*.

(∵ $BG = 4\sqrt{2}$ and Δ *BDG* is a regular triangle.

∴ The area of Δ *BDG* is $\frac{\sqrt{3}}{4} \cdot \left(4\sqrt{2}\right)^2 = \frac{\sqrt{3}}{4} \cdot 16 \cdot 2 = 8\sqrt{3}$)

(3)

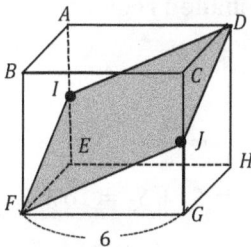

For a cube with sides of length 6,

points *I* and *J* are midpoints of $\overline{AE}$ and $\overline{CG}$.

Find the area of ▱ *IFJD*.

(∵ Since ▱ *IFJD* is a rhombus, $IJ = 6\sqrt{2}$.

The length of the diagonal is $6\sqrt{3}$.

Therefore, the area of ▱ *IFJD* is $\frac{1}{2} \cdot 6\sqrt{2} \cdot 6\sqrt{3} = 18\sqrt{6}$.)

(4)

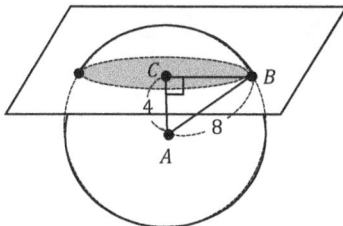

$AC = 4$, $AB = 8$.

Find the area of the cross section.

(∵ $CB^2 = 8^2 - 4^2 = 64 - 16 = 48 = 16 \cdot 3$; $CB = 4\sqrt{3}$

Therefore, the area of the cross section is $\pi r^2 = \pi(4\sqrt{3})^2 = 48\,\pi$)

(5)

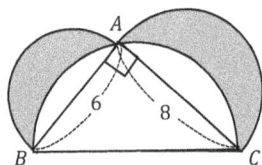

Find the area of the shaded regions.

(∵ The area of the shaded regions is the area of $\triangle ABC = \frac{1}{2} \cdot 6 \cdot 8 = 24$.)

(6)

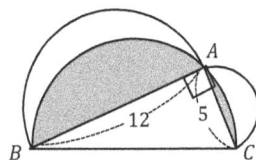

Find the area of the shaded regions.

(∵ $BC = 13$. The area of the shaded regions is $\frac{1}{2} \cdot \pi \left(\frac{13}{2}\right)^2 - \frac{1}{2} \cdot 5 \cdot 12 = \frac{169}{8}\pi - 30$)

(7)

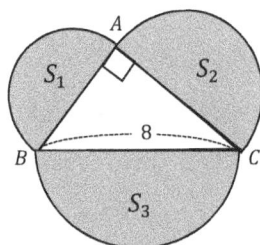

Find the area of the shaded regions.

(∵ $S_1 + S_2 = S_3$.

$S_3 = \frac{1}{2} \cdot \pi \left(\frac{8}{2}\right)^2 = \frac{1}{2} \cdot \pi \cdot \frac{64}{4} = \frac{64}{8}\pi = 8\pi$

Therefore, $S_1 + S_2 + S_3 = S_3 + S_3 = 16\pi$)

(8)

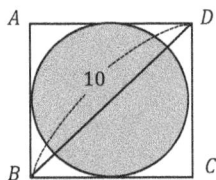

For a square $\square\,ABCD$ with a diagonal of length 10, find the area of the inscribed circle.

(∵ Let r be the radius of the inscribed circle. Then,

$CD = 2r$. So, $\sqrt{2} \cdot 2r = 10$.; $r = \frac{10}{2\sqrt{2}} = \frac{5\sqrt{2}}{2}$

Therefore, the area of the inscribed circle is $\pi \left(\frac{5\sqrt{2}}{2}\right)^2 = \pi \frac{25 \cdot 2}{4} = \frac{25}{2}\pi$)

#12 Find the volume of each figure.

(1) 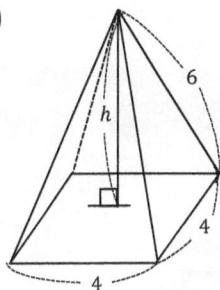 **(2)** 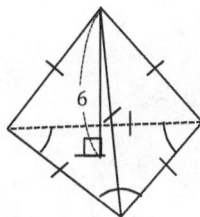 **(3)** 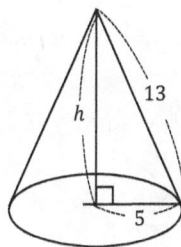 **(4)**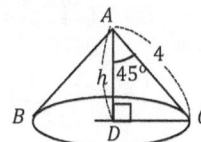

$(\because$ (1) $h = \sqrt{6^2 - \left(2\sqrt{2}\right)^2} = \sqrt{36 - 8} = \sqrt{28} = 2\sqrt{7}$

$V = \frac{1}{3} \cdot 4^2 \cdot 2\sqrt{7} = \frac{32\sqrt{7}}{3}$

(2) $h = \frac{\sqrt{6}}{3} a = 6$ So, the length of a side is $3\sqrt{6}$.

Therefore, $V = \frac{\sqrt{2}}{12} \cdot \left(3\sqrt{6}\right)^3 = \frac{27\sqrt{12}}{2} = 27\sqrt{3}$

(3) $h = 12$

Therefore, $V = \frac{1}{3} \cdot \pi \cdot 5^2 \cdot 12 = 100\,\pi$

(4) Since $\triangle ACD$ is a 45°- 45°- 90° triangle, $h : 4 = 1 : \sqrt{2}$; $h = \frac{4}{\sqrt{2}} = \frac{4\sqrt{2}}{2} = 2\sqrt{2}$

Since $h : DC = 1 : 1$, $h = DC = 2\sqrt{2}$

Therefore, $V = \frac{1}{3} \cdot \pi \cdot \left(2\sqrt{2}\right)^2 \cdot 2\sqrt{2} = \frac{16\sqrt{2}}{3}\,\pi$)

Solutions for Chapter 10

#1 Find the trigonometric ratios of $\angle x$.

(1)

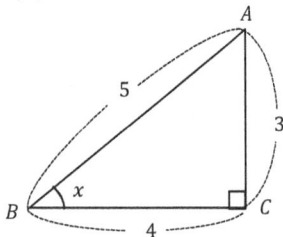

(2)

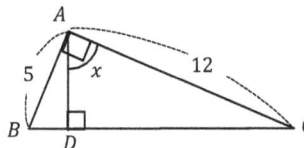

$\therefore$ (1) $\sin x = \frac{3}{5}$, $\cos x = \frac{4}{5}$, and $\tan x = \frac{3}{4}$

(2)

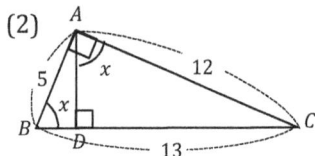

$\sin x = \frac{12}{13}$, $\cos x = \frac{5}{13}$, and $\tan x = \frac{12}{5}$

#2 Find the value of the following :

(1) $\sin 30° + \cos 30° = \frac{1}{2} + \frac{\sqrt{3}}{2} = \frac{1+\sqrt{3}}{2}$

(2) $\sin 45° + \cos 60° = \frac{\sqrt{2}}{2} + \frac{1}{2} = \frac{1+\sqrt{2}}{2}$

(3) $2\tan 30° \times \sin 60° \times \cos 45° = 2 \cdot \frac{\sqrt{3}}{3} \cdot \frac{\sqrt{3}}{2} \cdot \frac{\sqrt{2}}{2} = \frac{3\sqrt{2}}{6} = \frac{\sqrt{2}}{2}$

(4) $\sin^2 60 + \cos^2 60 = (\frac{\sqrt{3}}{2})^2 + (\frac{1}{2})^2 = \frac{3}{4} + \frac{1}{4} = 1$

#3 Find the values of x and y for the following right triangles $\Delta\,ABC$.

(1)

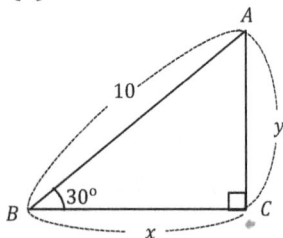

(2)

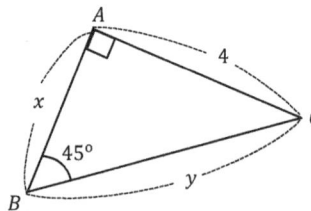

∵ (1) $\sin 30° = \frac{y}{10} = \frac{1}{2}$ ∴ $y = 5$

$\cos 30° = \frac{x}{10} = \frac{\sqrt{3}}{2}$ ∴ $x = 5\sqrt{3}$

(2) $\sin 45° = \frac{4}{y} = \frac{\sqrt{2}}{2}$ ∴ $y = \frac{8}{\sqrt{2}} = 4\sqrt{2}$

$\cos 45° = \frac{x}{y} = \frac{x}{4\sqrt{2}} = \frac{\sqrt{2}}{2}$ ∴ $2x = 8$; $x = 4$

(3)
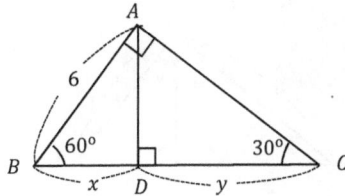

(4)
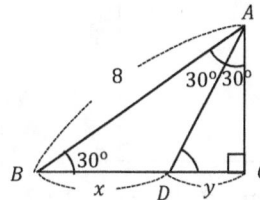

∵ (3) $\cos 60° = \frac{x}{6} = \frac{1}{2}$ ∴ $x = 3$

Since $\sin 60° = \frac{AD}{6} = \frac{\sqrt{3}}{2}$, $AD = 3\sqrt{3}$

$\tan 30° = \frac{AD}{y} = \frac{3\sqrt{3}}{y} = \frac{\sqrt{3}}{3}$ ∴ $y = \frac{9\sqrt{3}}{\sqrt{3}} = 9$

(4) Since $AC = 4$, $\tan 30° = \frac{y}{AC} = \frac{y}{4} = \frac{\sqrt{3}}{3}$ ∴ $y = \frac{4\sqrt{3}}{3}$

Since $BC = 4\sqrt{3}$, $x = 4\sqrt{3} - \frac{4\sqrt{3}}{3} = \frac{8\sqrt{3}}{3}$

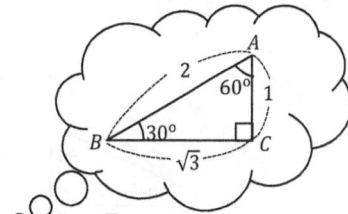

#4 For right triangles ΔABC with $m \angle C = 90°$, answer the following :

(1)
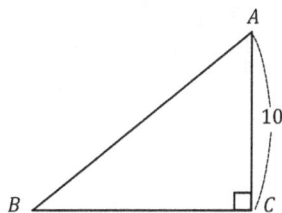

$\sin B = \frac{5}{13}$

Find the value of $\sin A$ and $\tan A$.

(2)
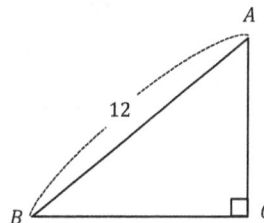

$\sin A = \frac{2}{3}$

Find the value of $\cos B \times \tan B$.

∵ (1)

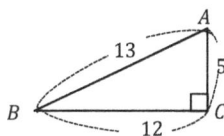 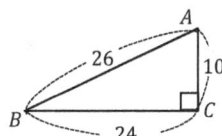

$\sin A = \frac{24}{26} = \frac{12}{13}$ and

$\tan A = \frac{24}{10} = \frac{12}{5}$

∵ (2)

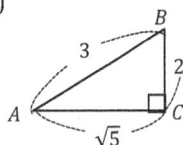

$\cos B = \frac{8}{12}$ and $\tan B = \frac{4\sqrt5}{8}$

So, $\cos B \times \tan B = \frac{8}{12} \cdot \frac{4\sqrt5}{8} = \frac{\sqrt5}{3}$

(3)

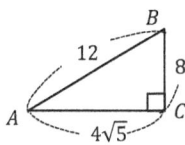

$m \angle A : m \angle B : m \angle C = 1 : 2 : 3$

Find the value of $\sin B \times \cos B \times \tan B$.

(4)

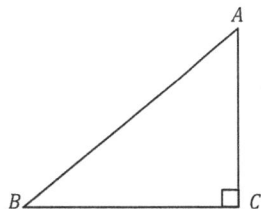

Find the value of $\sin x + \cos y$.

$m \angle A + m \angle B + m \angle C = 180°$

Let $m \angle A = x$.

Then, $x + 2x + 3x = 180°$

So, $6x = 180°$; $x = 30°$

Therefore,

$\sin B \times \cos B \times \tan B$

$= \sin 60° \times \cos 60° \times \tan 60°$

$= \frac{\sqrt3}{2} \cdot \frac{1}{2} \cdot \sqrt3 = \frac{3}{4}$

∵ (3)

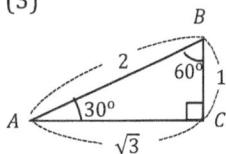

∵ (4)

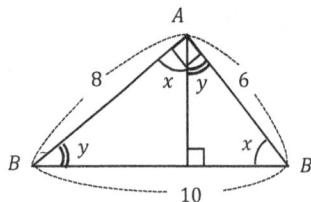

Since $\sin x = \frac{8}{10}$ and $\cos y = \frac{8}{10}$,

$\sin x + \cos y = \frac{8}{10} + \frac{8}{10} = \frac{16}{10} = \frac{8}{5}$

(5)

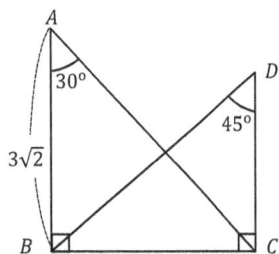

$m \angle C = m \angle B = 90°$

Find BD.

(6)

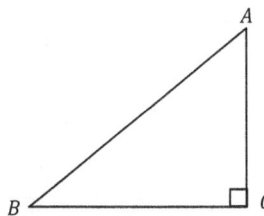

$\sin A = \frac{\sqrt7}{4}$

Find $\tan B$.

$\because$ (5) Since $\tan 30^\circ = \frac{BC}{3\sqrt{2}} = \frac{\sqrt{3}}{3}$, $BC = \frac{3\sqrt{6}}{3} = \sqrt{6}$

Since $\sin 45^\circ = \frac{BC}{BD} = \frac{\sqrt{6}}{BD} = \frac{\sqrt{2}}{2}$, $BD = \frac{2\sqrt{6}}{\sqrt{2}} = \frac{2\sqrt{12}}{2} = 2\sqrt{3}$

$\because$ (6)

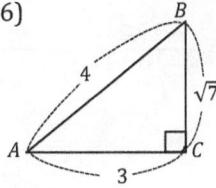

$AC = \sqrt{16 - 7} = \sqrt{9} = 3$

So, $\tan B = \frac{3}{\sqrt{7}} = \frac{3\sqrt{7}}{7}$

#5 **Find the value of x for the following triangles :**

(1)

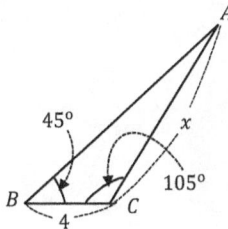

(2)

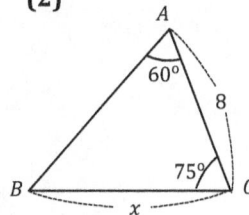

(3)

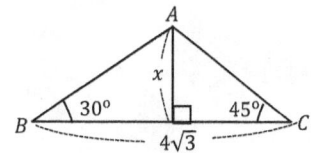

$\because$ (1)

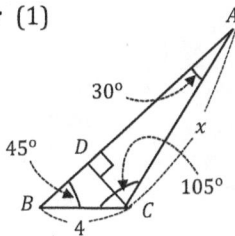

$\sin 45^\circ = \frac{CD}{4} = \frac{\sqrt{2}}{2}$ $\therefore CD = 2\sqrt{2}$

$\sin 30^\circ = \frac{CD}{x} = \frac{2\sqrt{2}}{x} = \frac{1}{2}$ $\therefore x = 4\sqrt{2}$

$\because$ (2)

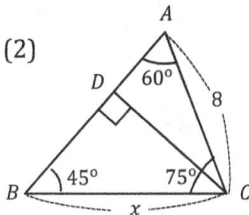

$\sin 60^\circ = \frac{CD}{8} = \frac{\sqrt{3}}{2}$ $\therefore CD = 4\sqrt{3}$

$\sin 45^\circ = \frac{4\sqrt{3}}{x} = \frac{\sqrt{2}}{2}$ $\therefore x = \frac{8\sqrt{3}}{\sqrt{2}} = \frac{8\sqrt{6}}{2} = 4\sqrt{6}$

∵ (3)

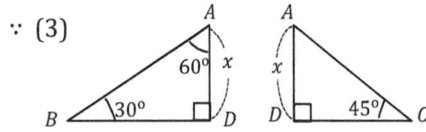

$\tan 60° = \frac{BD}{x} = \sqrt{3} \quad \therefore BD = \sqrt{3}\,x$

$\tan 45° = \frac{x}{DC} = 1 \quad \therefore DC = x$

Since $BD + DC = BC$, $\sqrt{3}\,x + x = 4\sqrt{3}$

So, $(1 + \sqrt{3})x = 4\sqrt{3}$

Therefore,

$$x = \frac{4\sqrt{3}}{1+\sqrt{3}} = \frac{4\sqrt{3}\,(\sqrt{3}-1)}{(1+\sqrt{3})\,(\sqrt{3}-1)} = \frac{4\sqrt{3}\,(\sqrt{3}-1)}{3-1} = 2\sqrt{3}\,(\sqrt{3}-1).$$

(4)

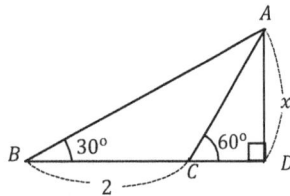

(5)

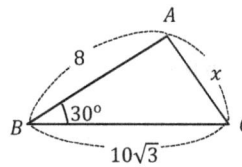

∵ (4) $\tan 30° = \frac{x}{BD} = \frac{\sqrt{3}}{3} \quad \therefore BD = \frac{3x}{\sqrt{3}} = \sqrt{3}\,x$

$\tan 60° = \frac{x}{CD} = \sqrt{3} \quad \therefore CD = \frac{x}{\sqrt{3}} = \frac{\sqrt{3}\,x}{3}$

Since $BC + CD = BD$, $2 + \frac{\sqrt{3}\,x}{3} = \sqrt{3}\,x$

Therefore, $\sqrt{3}\,x\left(1 - \frac{1}{3}\right) = 2$; $\sqrt{3}\,x = 2 \cdot \frac{3}{2} = 3 \quad \therefore x = \sqrt{3}$

∵ (5)

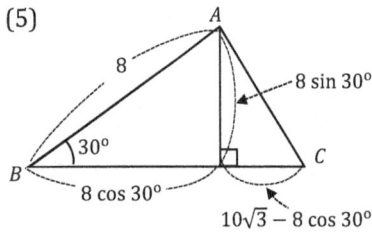

$x = AC$

$= \sqrt{(8\sin 30°)^2 + (10\sqrt{3} - 8\cos 30°)^2}$

$= \sqrt{(8 \cdot \frac{1}{2})^2 + (10\sqrt{3} - 8 \cdot \frac{\sqrt{3}}{2})^2}$

$= \sqrt{16 + 36 \cdot 3} = \sqrt{124} = 2\sqrt{31}$

#6 Find the areas of the following :

(1)

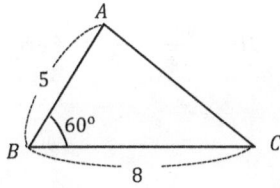

(2)

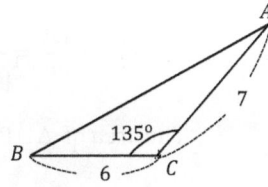

(3)

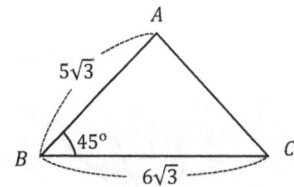

$\therefore$ (1) $S = \frac{1}{2} \cdot 5 \cdot 8 \cdot \sin 60° = 20 \cdot \frac{\sqrt{3}}{2} = 10\sqrt{3}$

(2) $S = \frac{1}{2} \cdot 6 \cdot 7 \cdot \sin(180° - 135°) = 21 \cdot \sin 45° = 21 \cdot \frac{\sqrt{2}}{2} = \frac{21\sqrt{2}}{2}$

(3) $S = \frac{1}{2} \cdot 5\sqrt{3} \cdot 6\sqrt{3} \cdot \sin 45° = 45 \cdot \frac{\sqrt{2}}{2} = \frac{45\sqrt{2}}{2}$

(4)

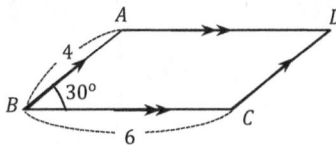

(5)

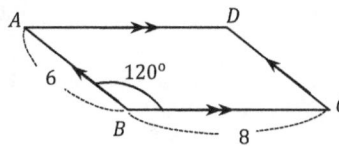

(6)

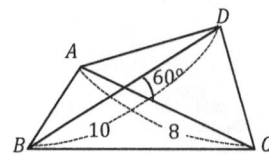

(4) $S = 4 \cdot 6 \cdot \sin 30° = 24 \cdot \frac{1}{2} = 12$

(5) $S = 6 \cdot 8 \cdot \sin(180° - 120°) = 48 \cdot \sin 60° = 48 \cdot \frac{\sqrt{3}}{2} = 24\sqrt{3}$

(6) $S = \frac{1}{2} \cdot 8 \cdot 10 \cdot \sin 60° = 40 \cdot \frac{\sqrt{3}}{2} = 20\sqrt{3}$

(7)

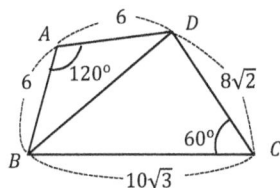

(8)

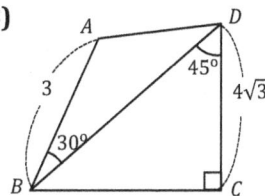

(7) $S = \frac{1}{2} \cdot 6 \cdot 6 \cdot \sin(180° - 120°) + \frac{1}{2} \cdot 10\sqrt{3} \cdot 8\sqrt{2} \cdot \sin 60°$

$= 18 \cdot \sin 60° + 40\sqrt{6} \cdot \sin 60° = 18 \cdot \frac{\sqrt{3}}{2} + 40\sqrt{6} \cdot \frac{\sqrt{3}}{2} = 9\sqrt{3} + 20\sqrt{18} = 9\sqrt{3} + 60\sqrt{2}$

(8)

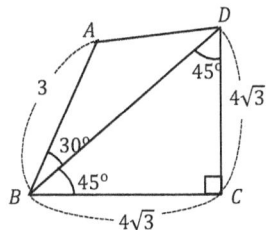

The area of $\triangle DBC$ is $\frac{1}{2} \cdot 4\sqrt{3} \cdot 4\sqrt{3} = 24$.

$\text{Sin } 45° = \frac{4\sqrt{3}}{BD} = \frac{\sqrt{2}}{2}$ $\therefore BD = \frac{8\sqrt{3}}{\sqrt{2}} = \frac{8\sqrt{6}}{2} = 4\sqrt{6}$

The area of $\triangle ABD$ is

$\frac{1}{2} \cdot 3 \cdot 4\sqrt{6} \cdot \text{Sin } 30° = \frac{1}{2} \cdot 3 \cdot 4\sqrt{6} \cdot \frac{1}{2} = 3\sqrt{6}$

Therefore, the area of $\square ABCD$ is $24 + 3\sqrt{6}$.

Index

A

B

C

D

O

Oblique cylinder, 161

Oblique prism, 156

Obtuse, 199

Obtuse angle, 29, 225, 227

Opposite , 218

Opposite angles, 66

P

Parallel, 178

Parallel lines, 32, 175

Parallelogram, 63, 67, 142, 185, 227

Perimeter, 51, 139, 144, 183

Perimeter of an isosceles triangle, 91

Perpendicular , 131

Perpendicular bisector, 27, 124, 130

Perpendicular segment, 27

Plane, 23

Polygons, 51, 144, 203

Polynomial region, 138

Positive number, 20

Prism, 156

Proof, 196

Proportionality theorem, 175, 176

Pyramid, 158

Pythagorean Theorem, 196, 219, 223

Q

Quadrilateral, 62, 203

R

Radius, 84

Ratio of circumference, 145

Rational number, 20

Ray, 22

Real number, 20

Rectangle, 63, 71, 138, 203

Regular polygon, 52, 63, 144

Regular prism, 156

Regular pyramid, 158

Regular rectangular prism, 205

Regular rectangular pyramid, 205

Regular triangle, 204

Regular triangular pyramid, 206

Relationship of the sides and angles, 199

Remote interior angles, 54

RHA correspondence, 47

Rhombus, 70, 142

RHS correspondence, 47

Right, 199, 42

Right angle, 29

Right cylinder, 161

Right prism, 156

Right rectangular prism, 205

Right triangle, 47, 223, 199

S

SAA correspondence and postulate, 44

SAS correspondence and postulate, 43

SAS similarity theorem, 163

Scalene, 42

Secant, 84, 99, 101

Sector, 147

Segment, 22

Segment proportionality theorem, 178

Side, 28

Similar, 170, 173

Similar triangles, 180

Similarities in right triangles, 174

Similarity, 49, 170, 199

Similar triangles, 183